Assessing Revolutionary and Insurgent Strategies

UNCONVENTIONAL WARFARE CASE STUDY: THE RHODESIAN INSURGENCY AND THE ROLE OF EXTERNAL SUPPORT: 1961–1979

Paul J. Tompkins Jr., USASOC Project Lead

Mark Grdovic, Brad Gutierrez, Ted Pilkulsky, Guillermo Pinczuk, and Bill Riggs, Contributing Authors

United States Army Special Operations Command
and
The Johns Hopkins University Applied Physics Laboratory
National Security Analysis Department

This publication is a work of the United States Government in accordance with Title 17, United States Code, sections 101 and 105.

Published by:

The United States Army Special Operations Command

Fort Bragg, North Carolina

Reproduction in whole or in part is permitted for any purpose of the United States government. Nonmateriel research on special warfare is performed in support of the requirements stated by the United States Army Special Operations Command, Department of the Army. This research is accomplished at the Johns Hopkins University Applied Physics Laboratory by the National Security Analysis Department, a nongovernmental agency operating under the supervision of the USASOC Sensitive Activities Division, Department of the Army.

The analysis and the opinions expressed within this document are solely those of the authors and do not necessarily reflect the positions of the US Army or the Johns Hopkins University Applied Physics Laboratory.

Comments correcting errors of fact and opinion, filling or indicating gaps of information, and suggesting other changes that may be appropriate should be addressed to:

United States Army Special Operations Command

G-3X, Sensitive Activities Division

2929 Desert Storm Drive

Fort Bragg, NC 28310

All ARIS products are available from USASOC at www.soc.mil under the ARIS link.

Published by Conflict Research Group.

First published by USASOC in 2019

CONFLICT
RESEARCH
GROUP

ASSESSING REVOLUTIONARY AND INSURGENT STRATEGIES

The Assessing Revolutionary and Insurgent Strategies (ARIS) series consists of a set of case studies and research conducted for the US Army Special Operations Command by the National Security Analysis Department of the Johns Hopkins University Applied Physics Laboratory.

The purpose of the ARIS series is to produce a collection of academically rigorous yet operationally relevant research materials to develop and illustrate a common understanding of insurgency and revolution. This research, intended to form a bedrock body of knowledge for members of the Special Forces, will allow users to distill vast amounts of material from a wide array of campaigns and extract relevant lessons, thereby enabling the development of future doctrine, professional education, and training.

From its inception, ARIS has been focused on exploring historical and current revolutions and insurgencies for the purpose of identifying emerging trends in operational designs and patterns. ARIS encompasses research and studies on the general characteristics of revolutionary movements and insurgencies and examines unique adaptations by specific organizations or groups to overcome various environmental and contextual challenges.

The ARIS series follows in the tradition of research conducted by the Special Operations Research Office (SORO) of American University in the 1950s and 1960s, by adding new research to that body of work and in several instances releasing updated editions of original SORO studies.

VOLUMES IN THE ARIS SERIES

Casebook on Insurgency and Revolutionary Warfare, Volume I: 1927–1962 (Rev. Ed.)
Casebook on Insurgency and Revolutionary Warfare, Volume II: 1962–2009
Case Studies in Insurgency and Revolutionary Warfare: Algeria 1954–1962 (pub. 1963)
Case Studies in Insurgency and Revolutionary Warfare—Colombia (1964–2009)
Case Studies in Insurgency and Revolutionary Warfare: Cuba 1953–1959 (pub. 1963)
Case Study in Guerrilla War: Greece During World War II (pub. 1961)
Case Studies in Insurgency and Revolutionary Warfare: Guatemala 1944–1954 (pub. 1964)
Case Studies in Insurgency and Revolutionary Warfare—Palestine Series
Case Studies in Insurgency and Revolutionary Warfare—Sri Lanka (1976–2009)
Unconventional Warfare Case Study: The Relationship between Iran and Lebanese Hizbollah
Unconventional Warfare Case Study: The Rhodesian Insurgency and the Role of External Support: 1961–1979
Human Factors Considerations of Undergrounds in Insurgencies (2nd Ed.)
Irregular Warfare Annotated Bibliography
Legal Implications of the Status of Persons in Resistance
Narratives and Competing Messages
Special Topics in Irregular Warfare: Understanding Resistance
Threshold of Violence
Undergrounds in Insurgent, Revolutionary, and Resistance Warfare (2nd Ed.)

SORO STUDIES

Case Studies in Insurgency and Revolutionary Warfare: Vietnam 1941–1954 (pub. 1964)

TABLE OF CONTENTS

LIST OF ILLUSTRATIONS

LIST OF TABLES

CHAPTER 1.
INTRODUCTION

EXECUTIVE SUMMARY

The political and military contest black nationalists in Rhodesia waged against the white minority governments of Winston Fields (1962–1964) and Ian Smith (1964–1979) provides an interesting case study through which to examine the dynamics of an insurgency that had the training, support, and advisory assistance of external sponsors. The Rhodesian conflict is a unique case because the two external sponsors, the Soviet Union and China, provided support to two competing insurgent groups, the Zimbabwe African People's Union (ZAPU) and the Zimbabwe African National Union (ZANU), respectively. This dichotomy allows for the comparison and contrast of two approaches to unconventional warfare and the corresponding strategies and tactics implemented by the recipients of that support.

Additionally, the Rhodesian conflict was further complicated by the anomaly of Western and Communist powers ostensibly aligned in goal, if not in technique. While the Soviet Union and China were supporting their clients in undertaking an insurrection against the white regime, the United States, Great Britain, and other Western powers were imposing economic sanctions, to varying degrees of success, in hopes of achieving a political solution to the ultimate goal of expediting the process of handing over rule to the majority black African population.

A significant body of literature written during and after the Rhodesian conflict details the counterinsurgency efforts of the various elements of the Rhodesian Security Forces (RSF). This case study focuses on the insurgents, their external supporters, how the former used that external support, and the effectiveness of insurgent activities. In doing so, this study highlights a number of lessons for military professionals regarding unconventional warfare and support to partner insurgent forces, including

1. the importance of ensuring that external support is tailored to the requirements and environment of the insurgent forces and their battlefields;

2. the criticality of realistic and appropriate insurgent strategies if external support is to be effective;

3

3. the importance of understanding the role of local "structural" conditions that impact an insurgency (which in this case refers to a number of exogenous factors, including geography and the human terrain, as well as the status of independence movements in surrounding territories, such as Angola and Mozambique, and the interests of newly independent states in southern and eastern Africa in the Rhodesian conflict); and

4. perhaps most importantly in the Rhodesian case, the requirement for unity of effort among insurgent groups (arguably, if the two nationalist groups could have overcome their differences and formed a truly singular insurgent movement, they may have attained success much sooner).

INTRODUCTION

For nearly twenty years during the Cold War, black Rhodesian nationalists waged a political and military campaign to oust the white minority regime of Ian Smith and his Rhodesian Front (RF) government. It was a campaign characterized by competing insurgent groups relying on the support of ideologically similar, yet politically competing, external powers. Additionally, it was conducted under the watchful eye of a regional intergovernmental organization, the Organization of African Unity (OAU), keenly interested in trying to orchestrate the ultimate outcome. The lessons to be drawn from this case are many but essentially focus on the need to attain unity of effort and the importance of tying military strategy to political objectives.

The goal of the Rhodesian insurgency was to establish black African majority rule in the former British colony, thereby deposing the white government that had been in control since Britain had granted self-rule to the colony of Southern Rhodesia in 1923. Although Britain had played a key role in negotiating the transfer of power from white minority rule in Rhodesia's neighboring countries of Zambia (formerly Northern Rhodesia) and Malawi (formerly Nyasaland), culminating in independence for the two countries in 1964, Ian Smith's Unilateral Declaration of Independence (UDI) from Britain in 1965 diminished the likelihood of a negotiated transfer from minority to majority rule in Rhodesia. Despite the international sanctions placed on Rhodesia and their economic consequences, the Smith regime waged a determined counterinsurgency campaign late into the 1970s. Not until the apartheid regime in South Africa was pressured to curtail its support for Smith in 1976 did the Rhodesian government acknowledge that change was inevitable and the ninety-year reign of the white settler in the colony founded by Cecil Rhodes come to an end.

What ultimately became a negotiated settlement engineered by the British government and signed at Lancaster House in London in 1979, creating the country of Zimbabwe, was the result of a political and military stalemate. In their earlier attempts, British and American interlocutors could not meet all the Rhodesian demands for a peaceful transfer of power. The two major insurgent groups, ZANU and ZAPU, could not agree on military or political strategies for a united front. Despite the lack of military evidence to support their confidence, both the Rhodesian government and the insurgents felt they could still win a military victory. Finally, pressure from a variety of international interlocutors, including Mozambique, Zambia, South Africa, the United States, Great Britain, the Soviet Union, and the OAU, brought the parties to Lancaster House. The result was an open election in which the black majority elected Robert Mugabe, the head of ZANU, to power. Soon after Mugabe assumed power, the country adopted the name Zimbabwe in place of Rhodesia.

The focus of this study is the insurgents, their supporters, and the strategies and tactics both employed. A number of factors make the Rhodesian case unique among the majority of African independence movements in the 1950s and early 1960s. First, the Rhodesian insurgency was waged against a white government that did not enjoy widespread support from the non-Communist countries in Europe and North America. After UDI the Ian Smith-led regime was, in fact, an international leper subject to wide-ranging, albeit haphazardly enforced, economic sanctions. Second, the insurgency was relatively late in gaining its momentum. While many of its African neighbors attained independence in the first half of the 1960s as part of the decade of African independence, the insurgency in Rhodesia did not hit its stride until the early 1970s when armed guerrilla action became a staple of the insurgency effort. Even then, the infighting between ZANU and ZAPU leaders made the insurgent efforts somewhat sporadic and without consistency of message or effort.

Third, the Rhodesian insurgency was contested in an environment so politically charged that it often appeared leaderless. Soviet and Chinese sponsors had their own political agendas that may or may not have always been beneficial to their ZAPU and ZANU beneficiaries. The OAU sought to mask the internal conflicts between ZAPU and ZANU to help quell fears that neither of the insurgent factions would be able to effectively govern a new majority-led Zimbabwe. Neighboring countries, particularly Zambia and Mozambique, grew to feel that the dissension between ZANU and ZAPU was unnecessarily extending the insurgency, causing undue damage in their own countries from Rhodesian counterinsurgency raids on camps used by the military arms of ZANU and ZAPU. Finally, white Rhodesia's largest supporter, apartheid South Africa, faced increasing international pressure to curtail its military and economic support to the Smith regime, lest it also become the target of international pressure to end white minority rule.

As mentioned above, the lead external supporters for the insurgent groups in Rhodesia were the Soviets and the Chinese, who supported ZAPU and ZANU, respectively. This support was provided through the provision of arms as well as training in Africa, the Soviet Union, China, and North Korea. As summarized in a 1975 Rhodesian Foreign Ministry report entitled *Communist Support and Assistance to Nationalist Political Groups in Rhodesia*, Soviet training for ZAPU forces was conducted in Simferopol, Odessa (Ukraine), and in Moscow. The training consisted of four main types: paramilitary, military engineering, radio, and intelligence.[1] Other Soviet-based paramilitary training was provided in Bulgaria, North Korea, and Egypt. Initial Chinese training for ZANU forces was conducted in military bases near Peking and Nanking.[2] As the liberation struggle continued, Chinese instructors conducted training in Tanzania.[3]

It is important to note at this point that the relationship between the two Communist behemoths and the African continent can in no way be simplified to that of two apples from different branches of the same Communist, anticapitalist tree. The czarist predecessors of the Soviet regime had relationships with African states dating back to the opening of the Suez Canal in 1869, at which time Russia gave military support to Ethiopia for protection against perceived British threats.[4] In the late 1920s, Soviet interest in Africa increased as the leadership tried to exploit emerging nationalist movements in Africa, seeing the colonies as the weak link in the imperialists' chain of global influence.[5]

The rise of Stalin and the coming of the Second World War diverted Soviet attention. That attention was not refocused on Africa until the 1955 Afro-Asian Conference in Bandung, Indonesia, after which China also began to extend its reach to the African continent.

During the conflict, the Soviet Union viewed its involvement in Africa as part of its competition with the United States for global influence. By comparison, the People's Republic of China (PRC) viewed itself as a national liberation partner for Africans fighting their own struggles for independence and as a champion for the developing world. It also saw the economic potential in partnerships with emerging African states.[6] Additionally, both countries viewed themselves in competition with one another over leadership of the communist world and to be seen as the key sponsor of revolutionary movements in the developing world. In the case of support for resistance movement in Rhodesia, while the Soviets focused on conventional strategies as well as guerrilla strategies akin to partisan tactics employed against invading Nazi forces during World War II, the Chinese emphasized a "people's war" strategy. As will be discussed in greater detail in later chapters of this study, the employment of specific military strategies had an important impact on the ultimate political goals of the insurgents, which were to assume power and achieve majority rule.

THE PURPOSE AND METHODOLOGY OF THIS CASE STUDY

The purpose of this study is to examine the insurrections waged by ZANU and its military arm, the Zimbabwe African National Liberation Army (ZANLA), and ZAPU and its military forces, the Zimbabwe People's Revolutionary Army (ZIPRA), through the support they received from the Soviet Union and China, as well as from the neighboring states of Zambia, Mozambique, and Tanzania. Through secondary sources covering a wide spectrum of intellectual inquiry including political science, history, military doctrine and tactical analysis, news accounts from the insurgency period, and documented interviews with insurgents, this case study highlights not only the actions and events of the period but also the lessons learned from the actions of the insurgents and their external supporters. Ideally these lessons will provide context and perspective regarding the application of unconventional

7

warfare that may inform the decision-making process of other military and political leaders as they consider options for future campaigns.

This case study is organized into five main chapters following this one. Chapter 2 discusses the two main external supporters, particularly their views on Africa and, more broadly, on national liberation movements in the developing world. Chapter 3 discusses the social, historical, political, and economic context that shaped the conflict. Chapter 4 provides an overview of the insurrection in Rhodesia and how it developed into two separate insurgencies. This chapter also looks into the development of the respective strategies and the prosecution of their campaigns. Chapter 5 discusses how the two external sponsors provided support to their clients, with emphasis on the institutions and individuals responsible for the relationships and the nature of the training provided. Finally, the conclusion discusses the impact of the military strategies adopted by ZANU and ZAPU on the ultimate political objectives of seizing power and attaining majority rule.

NOTES

[1] Rhodesian Ministry of Foreign Affairs, *Communist Support and Assistance to Nationalist Political Groups in Rhodesia* (Rhodesia: Information Section, Ministry of Foreign Affairs, November 28, 1975), 4, posted on *Rhodesia and South Africa: Military History* (blog), accessed August 22, 2014, http://www.rhodesia.nl/commsupp.htm.

[2] Ibid.

[3] Ibid.

[4] John Barratt, *The Soviet Union and Southern Africa* (Braamfontein: The South African Institute of International Affairs, 1981), 2.

[5] Ibid., 3.

[6] Victor Ojakorotu and Ayo Whehto, "Sino-African Relations: The Cold War Years and After," *Asia Journal of Global Studies* 2, no. 2 (2008): 35–36.

CHAPTER 2.
THE EXTERNAL SPONSORS

DESCRIPTION AND STRATEGIC GOALS

In the 1960s and 1970s both the Zimbabwe African People's Union (ZAPU) and Zimbabwe African National Union (ZANU) waged insurgencies within an extraordinarily complex international environment. The landscape featured newly independent states in the northern part of sub-Saharan Africa eager to see an end to white minority rule in southern Africa. Additionally it included anticolonial insurgencies raging in Lusophone Africa, particularly to the east and west in Mozambique and Angola, respectively, and a regional hegemon (South Africa) committed to apartheid and white minority rule and seeking to roll back the revolutionary tide in southern Africa. And, most importantly, the environment included the participation of the two predominant Communist powers, the Soviet Union and the People's Republic of China (PRC), both of which were born in revolution and sought to overthrow the existing international system—thus making them, from the perspective of the United States, "rogue" states par excellence, but who were embroiled in a bitter dispute with each other.

As noted in the previous chapter, czarist Russia was not absent from Africa, and Russian Marxists also took an early interest in the continent, as the very first issue of the Marxist paper *Iskra*, which Lenin began editing in 1900, mentioned South Africa twice.[1] Earlier, some of the founding fathers of communism also commented on Africa. Karl Marx "condemned slavery and accused eighteenth century English capitalists of building their prosperity upon the ill-gotten profits of the slave trade."[2] For the Bolsheviks, Africa played a critical role in preventing the collapse of decadent European capitalism, as noted by Wilson:[3]

> For one thing, they [the Bolsheviks] readily perceived a strategic connection between the economic health of the capitalist system and the expanding scope of its African enterprises. European capitalism may have been in the process of decay, but its life could well be prolonged by the influx of African raw materials and by reliance upon "super profits" derived from the exploitation of colonial labor. Lenin was particularly concerned about the apparent ability of Cecil Rhodes and other "social chauvinists" to use such resources to bribe the European proletariat and thereby to stifle its revolutionary energies.

11

The international system and Russia's relationship with countries in the developing world experienced an inflection point with the Bolshevik revolution in 1917, which was perhaps telegraphed by Leon Trotsky's pithy statement upon assuming the position of the People's Commissar of Foreign Affairs: "I will issue a few revolutionary proclamations to the people of the world and then shut up shop."[4, a] Additionally, the Bolshevik revolutionaries believed that the revolution had to be exported to protect it at home. Jacobson noted that:[7]

> The Bolsheviks had come to power with two central expectations. They believed, first of all, that the imperialists would attempt to overthrow the revolution in Russia and that, with their combined forces, they were capable of doing so. The revolution in Russia would therefore not be secure until the threat of imperialist intervention had been eliminated by the spread of proletarian revolution to several, if not all, of the major powers of Europe. Second, they expected that the Russian Revolution would detonate a chain reaction of socialist revolutions that would spread throughout Europe and the world in a single movement, putting an end to socialist-capitalist opposition and rendering nations and national institutions obsolete, thus obviating the need for conventional interstate relations.

---

[a] Consistent with this theme, Trotsky was also (naively) critical of what he saw as the implements of capitalist diplomacy, such as secret treaties. As People's Commissar of Foreign Affairs, he published various secret treaties between Tsarist Russia and European countries that involved the disposition of territory. In a November 1917 article in *Izvestiia*, he noted that "In undertaking the publication of the secret diplomatic documents relating to the foreign diplomacy of the tsarist and the bourgeois coalition governments . . . we fulfill an obligation which our party assumed when it was the party of opposition. Secret diplomacy is a necessary weapon in the hands of the propertied minority which is compelled to deceive the majority in order to make the latter serve its interests. Imperialism, with its world-wide plans of annexation, its rapacious alliances and machinations, has developed the system of secret diplomacy to the highest degree. . . . The Russian people as well as the other peoples of Europe and those of the rest of the world should be given the documentary evidence of the plans which the financiers and industrialists, together with their parliamentary and diplomatic agents, were secretly scheming. . . . The government of workers and peasants abolished secret diplomacy with its intrigues, ciphers, and lies. We have nothing to hide. . . . We desire a speedy abolition of the supremacy of capital."[5] For Trotsky, any complaints from perplexed European ambassadors could be safely ignored, as they represented governments that would soon join the ash heap of history. See also Ulam's *Expansion and Coexistence*.[6]

> None of them were certain how long the entire process
> would take, but they were convinced that the October
> Revolution could not survive in isolation.

The problem, though, as noted earlier, is that colonial profits, in the opinion of Marxist writers, prevented the spread of revolution to the European continent. Interestingly, both Marx and Friedrich Engels, coauthors of the 1848 work *The Communist Manifesto*, were silent on the appropriate foreign policy of socialist states.[8] That topic was addressed by Vladimir Ilyich Ulyanov ("Lenin"). Africa, and the developing world more broadly, played an important role in Lenin's view of capitalist (and therefore Socialist) development. In his publication *Imperialism, the Highest Stage of Capitalism* (published in 1917), which Ulam noted is perhaps the foundational theoretical document explaining the sources of Soviet foreign policy,[9] Lenin noted that European colonial rivalry in Africa contributed to the outbreak of World War I, and he saw Africa as a breeding ground for future European imperialist wars which may drag in Russia.[10] Additionally, he established the earlier noted strategic connection between European capitalism and colonial profits by noting that the profits European colonial powers derived from their overseas possessions (including Africa) were used to provide a minimum level of social welfare benefits to the European proletariat, thereby prolonging decadent capitalism by dampening revolutionary fervor in Europe.[11, b] Hence, colonialism and imperialism were preventing the spread of communism, and revolutions in the colonial world could potentially accelerate the revolutionary process in Europe.[13]

Gorman noted that while Lenin's ideas on the role of imperialism in forestalling revolution were a source of "scriptural truth" that assumed "bible-like" status among contemporary African Socialists and Soviet theoreticians, Lenin was less clear on the process through which socialism would emerge in the colonial world.[14] Lenin did emphasize the prominent role to be played by the colonial world by noting:[15]

> The socialist revolution will not be solely, or chiefly,
> a struggle of the revolutionary proletarians in each
> country against their bourgeoisie—no, it will be a

---

[b] Several years later, Trotsky also noted the connection between the European proletariat and colonial subjects. In comments prepared for the First Comintern Congress in 1919, he stressed to the "colonial slaves of Africa and Asia" that "the hour of proletarian dictatorship in Europe will strike for you as the hour of your own emancipation."[12]

struggle of all the imperialist-oppressed colonies and countries, of all dependent countries, against international imperialism.

Lenin also believed that nationalist sentiment in the colonies would inevitably turn against capitalism and imperialism, but he was less clear on whether pre-capitalist societies could make the leap to socialism without first experiencing capitalism (and its inevitable contradictions) as an intermediary phase. In a report to the Second Congress of the Communist International in 1920, he vaguely noted that "backward peoples" could bypass the capitalist stage of development with the aid of countries that have made more revolutionary progress:[16]

> If the victorious revolutionary proletariat conducts systematic propaganda . . . and the Socialist governments come to their aid with all the means at their disposal—in that event it would be mistaken to assume that backward peoples must inevitably go through the capitalist stage of development.

The Communist International (later known as Comintern) itself was established as a Moscow-led association of national communist parties that sought the global spread and promotion of communism, and Rhodesia made a rhetorical appearance at the Third Comintern Congress in Moscow in May 1921. David Jones, one of the founders of the Communist Party of South Africa, appealed to the leadership of Comintern by noting that "Africa's hundred and fifty million natives are most easily accessible through the eight millions or so which comprise the native populations of South Africa and Rhodesia."[17] It was at the Fourth Comintern Congress in late 1922 (also in Moscow) where more explicit attention was focused on promoting communist revolution in Africa. Part of this focus was based on fears that European countries would mobilize African armies to fight wars, perhaps against Russia, or to suppress proletarian agitation in Europe. Jones himself noted "the time is pressing, the Negro armies of Imperialism are already on the Rhine."[18, c]

---

[c] Trotsky himself noted that "[t]he use of colored troops for imperialist war, and at the present time for the occupation of German territory, is a well thought out and carefully executed attempt of European capital...to raise armed forces . . . so that Capitalism may have mobilized, armed and disciplined African troops at its disposal against the revolutionary masses of Europe."[19] Communist fears may have been based on the fact that over

In terms of actual policy output, a "Negro Commission" of the Congress drafted a resolution titled "Thesis on the Negro Question," which was passed unanimously. It stated that "[t]he penetration and intensive colonization of regions inhabited by black races is becoming the last great problem on the solution of which the further development of capitalism itself depends."[21] It further noted that "the Negro problem has become a vital question of the world revolution," and "the cooperation of our oppressed black fellow-men is essential to the Proletarian Revolution and to the destruction of capitalist power."[22, d]

However, despite being home to what some regarded as the world's most exploited people,[24] Soviet activity in Africa at this time was quite limited owing to the European colonial powers' strong influence on the continent and the Soviet emphasis on developing socialism at home and protecting the Soviet Union from perceived threats from capitalist powers.[25] Soviet outreach at this time was largely limited to rhetorical support, including Comintern pronouncements—such as those from the Fourth Congress—expressing solidarity with colonial subjects in their struggle against colonialism, as well as the education of African students at the Communist University for Toilers of the East[e] and the establishment of relations with African Communist parties and like-minded groups.[27, f]

The existential crisis brought about by the Nazi invasion of the Soviet Union precluded any serious outreach and efforts targeting

---

181,000 soldiers were recruited from French West Africa for use in World War I, and in 1913, the Marxist writer Mikhail Pavlovich expressed concern that a proposed trans-Saharan railway would upset the European balance of power by enabling France to mobilize a large colonial African army to be used in the European theater.[20]

  [d] Interestingly, the Fourth Congress did not feature African representation, although it did include the presence of two African Americans.[23]

  [e] In the 1920s, African and African American students were enrolled in the Communist University of the Toilers of the East and the International Lenin School, both of which were located in Moscow. Students were given false identities during their stays in Moscow, and their curriculum included training in guerrilla warfare, espionage, and underground work. Yet students of African descent complained of racism in Moscow, and in 1932 they lodged a complaint against the "derogatory portrayal of Negroes in the cultural institutions of the Soviet Union" as "real monkeys."[26]

  [f] For instance, relations with the African National Congress (ANC) of South Africa date back to 1927, when Josiah Gumede, the president of the ANC, visited the Soviet Union on the tenth anniversary of the Bolshevik revolution. By that time relations with the South African Communist Party had already been established, and Gumede became head of the South African section of the League against Imperialism, a newly formed Soviet front organization.[28]

Africa. After the war, efforts were limited by Stalin's jaundiced view of nationalist leaders in the Third World who were not committed Communists. Soviet thinking at this time was dominated by the "two-camps" theory envisioning an epochal struggle between communism and capitalism that did not permit any middle ground.[29] Hence, Stalin viewed non-Communist leaders such as Nehru, Sukarno, and Gandhi as "lackeys of imperialism,"[30] while he viewed African nationalists as too "bourgeois," the "lickspittles and lackeys of colonialism and imperialism" and the "reserves" of imperialism.[31] Indeed, even a figure such as Kwame Nkrumah—who steered Ghana to independence from Great Britain and served as its first leader, and who accepted Lenin's analysis of imperialism yet regarded himself as an African Socialist rather than a Marxist-Leninist[32]—was regarded by the Soviets in 1954 as a shield "behind which the reality of British Imperialism and dominance conceals itself."[33]

Yet by the dawn of the liberation struggle in Rhodesia, the Soviets adopted a more flexible approach to Third World liberation movements and leaders who were not doctrinaire Marxist-Leninists. More specifically, Khrushchev acknowledged the emergence of a nonaligned or "neutralist" group of leaders who, by virtue of their emphasis on national liberation and anti-imperialism, were worthy of Soviet aid and support, even if such leaders were not hard-line communists. As long as they were genuinely "anti-imperialist" they merited Soviet support.[34] Following their leader's cue, Soviet Africanists at the time, such as Ivan Potekhin, no longer regarded the independence of African states as a farce but rather as an opportunity for the development of socialism and for the Soviet Union to make inroads.[35] Thus, it was not by coincidence that the first contacts between Joshua Nkomo, the leader of ZAPU, and Soviet authorities occurred while Khrushchev was leader of the Soviet Union.

This new, flexible approach Khrushchev adopted perhaps made more sense given similarities in Soviet and African leaders' conceptions of "liberation." As Kempton noted, both viewed national liberation as necessary to achieve political independence and as a means for reducing economic and political dependence on the capitalist West.[36] Hence, communist groups were sometimes at the forefront of efforts to achieve political independence and racial equality in Africa.

At the 1964 trial that sentenced him to life imprisonment, Nelson Mandela noted:[37]

> For many decades the Communists were the only political group in South Africa who were prepared to treat Africans as human beings and their equals; who were prepared to eat with us; talk with us, live with and work with us. Because of this, there are many Africans who today tend to equate communism with freedom.

Undoubtedly the Cold War was a battle of ideas, and Onslow suggested that socialism found a more receptive audience among African elites concerned with the political and economic modernization of their newly liberated societies.[38] Specifically, she noted:

> In Southern Africa, the Cold War also encompassed a battle of ideas about the appropriate path to progress and modernity. Here socialism appeared to provide to offer the path to true liberation through the transformation of the national political economy. Not only did it offer a unifying political creed that could transcend ethnic rivalries, inherited hierarchical structures and tensions within the artificial boundaries of the colonial territories. It appeared to offer a solution to the flawed economic legacies of colonialism—and a means to correct the asymmetry between the Western trading system and African underdeveloped economies. To Marxist theorists within, for example, ZANU, SWAPO [South West African People's Organization], the SAPC [South African Communist Party] and those in the ANC [African National Congress] hierarchy, its model of state-led development in societies in transition to urbanization and industrialization, possessed a "moral superiority." It also provided an ideological vehicle for the transformation of land ownership that did not perpetuate the domination of traditional power structures, or inherited colonial

patterns of land tenure—nor a reversion to peasant-based land ownership with its emphasis on subsistence agriculture.[g]

What were the main foreign policy goals motivating the Soviet Union to support liberation movements in southern Africa? Albright noted that the various colonial conflicts in southern Africa provided the Soviets with an opportunity to raise their profile and make a claim to global power status and, as part of this aim, to win acceptance for the Soviet political, economic, and military presence on the continent.[40] Of course, as the Cold War was raging at this time, the Soviet Union sought to weaken Western influence in the region, and the effort to promote the emergence of radical black governments in Rhodesia, Namibia, South Africa, and the former Portuguese colonies was seen as a means to increase Soviet influence in the region.[41,h] Lastly, the Sino-Soviet split and Chinese support for a variety of African liberation movements forced the Soviets' hand, as the Chinese presence, in addition to challenging the Soviet position as the leader of the global Communist movement, also threatened Soviet efforts to be seen as the

---

[g] She also noted the shortsightedness of attempting to import ideas regarding the organization of political and social life from Europe that grew out of very different social conditions than those that existed within Africa: "There seems to have been little awareness among the African elites at the start of the 1960s, of whether or not this European socio-economic model offered an appropriate answer to accelerated socio-economic development for newly independent African states, founded on very different social structures, patterns of population, with variably developed infrastructures and often gross inequalities of education and expertise, and the prevalence of rural economies with their massive disparities in land access and productivity."[39]

[h] Consistent with its competition with the United States, Soviet motivations were also based on realpolitik, as sub-Saharan Africa is home to huge mineral resources, especially metals used in the construction of advanced technology. Specifically, chromium and cobalt, which were used in the construction of jet engines and spacecraft, were key materials the United States imported, and the Soviets sought to restrict their supply to the West. As noted by Soviet Premier Leonid Brezhnev in a 1973 speech in Prague: "Our aim is to gain control of the two great treasure houses on which the West depends . . . the energy treasure house of the Persian Gulf and the mineral treasure house of central and southern Africa."[42] Additionally, Friedman noted that the Soviet Union sought to challenge Western shipping lanes off the African coast, which assumed greater importance after the closure of the Suez Canal during the 1967 conflict between Israel and its neighbors.[43] The canal remained closed for six years (until the end of the Yom Kippur War in 1973), and with the closure of the canal, all fuel tankers traveling from the Middle East to Europe were forced to travel around the whole of the African continent. Through greater influence in Angola and Mozambique, the Soviet Union sought to base an increased naval presence that could challenge those shipping lanes, and influence in Rhodesia and Zambia would serve as link between the two coasts.

chief benefactor of liberation and revolutionary movements throughout Africa.[44]

Sino-African relations span several millennia. The first Chinese contacts with Africa are believed to have occurred during the early Han dynasty under Emperor Wuti (140–87 BC), as a Chinese expedition sent in search of allies is believed to have reached Alexandria, in Egypt.[45] In the premodern era, they reached their apex with the voyages of Admiral Zheng He to the east coast of Africa in the early fifteenth century. Yet China's policy on Africa, and its foreign policy more generally, was placed on a radically new trajectory with the founding of the PRC in 1949. For Mao Tse-tung, the final defeat of the Kuomintang represented an opportunity not only to remake China's domestic society by turning it into a land of universal justice and equality. It also offered an opportunity to restore China's central position in world affairs, which in Chinese eyes was undermined by the military aggression, economic exploitation, and political subservience European imperial powers imposed on China in the nineteenth and early twentieth centuries.[46] More specifically, for Mao, the Chinese Revolution would restore China's rightful place in the world primarily through soft power because colonial subjects would inevitably look to the Chinese example to determine how best to throw off the yoke of colonial rule, thereby enabling China to stand at the forefront of a broader global movement championing national self-determination against imperialism and colonialism.[47, i]

In the late 1940s during the Chinese civil war, the relationship between Moscow and the Chinese Communist Party (CCP) was, according to one author, "close but not harmonious."[49] Yet for Mao the world was divided between "two camps" split between progressive and reactionary forces,[50] and for both ideological and security-related reasons[j]

---

[i] Brazinksy noted that "[t]he writings of Mao and other future CCP leaders exhibited a strong identification with Egyptians, Koreans, Indians, and other peoples who had lost their independence. They conceptualized themselves as part of what historian Michael H. Hunt has called a 'community of the weak and oppressed.' At the same time, Chinese revolutionaries saw this 'community' not only as a group that could sympathize with their plight but also as a venue to redeem China's status. They expected that, given China's historic centrality in world affairs, its revolution would naturally become an example for and influence on other revolutionary movements. China would gain prestige among other indigenous nationalists by helping them to wage revolution."[48]

[j] Jian noted that Mao and the top leaders of the CCP were deeply concerned about a possible American intervention in China during the later stages of the civil war, and

it was therefore necessary for China to "lean to one side" and support progressive forces throughout the world. In an article titled "On People's Democratic Dictatorship," Mao indicated that China must:[52]

> unite in a common struggle with those nations of the world that treat us as equal and unite with the peoples of all countries—that is, ally ourselves with the Soviet Union, with the People's Democratic Countries, and with the proletariat and the broad masses of the people in all other countries, and form an international front. . . . We must lean to one side.

For Mao and the CCP, the moral superiority of socialism over "imperialism" was eventually complemented by socialism's perceived technological and military superiority, which for them was demonstrated in August 1957 with the announcement that the Soviet Union had successfully launched an intercontinental ballistic missile and, in October and early November, its launching of two earth satellites. In a November 1957 speech in Moscow addressing a meeting of Communist and Workers' parties, Mao noted:[53]

> It is my opinion that the international situation has now reached a new turning point. There are two winds in the world today, the East Wind and the West Wind. There is a Chinese saying "Either the East Wind prevails over the West Wind or the West Wind prevails over the East Wind." I believe it is characteristic of the situation today that the East Wind is prevailing over the West Wind. That is to say, the forces of socialism have become overwhelmingly superior to the forces of imperialism.

Yet it is around this time that serious fissures emerged within the Sino-Soviet alliance. The proximate cause was Khrushchev's denunciation of Stalin in February 1956 during the Twentieth Congress of the Communist Party of the Soviet Union (CPSU). The denunciation came as a shock to Mao and the top leaders of the CCP, who believed

---

this fear was based on their perception that Western capitalist countries were inherently aggressive and evil. Given these beliefs, Jian noted that Mao found it necessary for China to side with other Socialist countries.[51]

that although Stalin had made a number of mistakes,[k] he should be regarded as a "great Marxist-Leninist revolutionary leader" whose leadership of the USSR provided China with a model for how to undertake its own political, social, and economic transformation.[55]

Additionally, with the death of Stalin, Mao believed that he, rather than Khrushchev, was uniquely qualified to lead the international Communist movement, particularly in the developing world.[56] At the very least, Mao sought parity for China with respect to the Soviet Union. Jian noted that Mao never enjoyed meeting Stalin and in particular resented the way Stalin treated him as an inferior "younger brother."[57] With Stalin's death, Mao sought to make China an equal partner with the Soviet Union within the international Communist movement, and he was hesitant to cede authority to what he and other members of the top CCP leadership regarded as a less sophisticated Soviet leadership that emerged after Stalin.[58]

Furthermore, although the leadership of both countries ascribed to a teleological view of history in which a Communist victory over capitalism was historically preordained (and that no permanent détente with capitalism was possible), significant differences emerged between the two Communist powers regarding the appropriate strategy to be adopted by the Socialist bloc in the developing world. Specifically, the Chinese believed that Soviet technological achievements in 1957 demonstrated that the socialist bloc was now more powerful than the West, and therefore China advocated a more activist approach to supporting national liberation groups in the developing world because the Socialist Bloc had nothing to fear from Western strategic power.[59] In particular, the Chinese believed that the United States could be effectively defeated in a Soviet first strike. For instance, one December 1958 article from the journal *Shih-chieh chih-shih* (*World Culture*) noted:[60]

---

k   After Japan's defeat in August 1945, Mao anticipated a renewal of the civil war with the Kuomintang (KMT), and he instructed his forces to be prepared for a renewal of hostilities. Furthermore, he expected Soviet backing. Instead, Stalin signed a treaty with Chiang Kai-shek, the leader of the KMT, in which the Soviets recognized Chiang as the leader of China's legal government. Stalin pressured the Chinese Communists to negotiate with Chiang while warning that the resumption of the civil war would be disastrous for China. Mao and other top Chinese Communist leaders viewed Stalin's policy as a betrayal. Additionally, Mao criticized Stalin's hesitance to sign a treaty with the PRC when Mao visited the Soviet Union in late 1949 to early 1950, with Stalin doing so only after Chinese troops came to the defense of North Korea.[54]

> The absolute superiority of the Soviet Union in inter-
> continental ballistic missiles has placed the striking
> capabilities of the United States . . . in an inferior posi-
> tion. The Soviet intercontinental ballistic missiles not
> only can reach any military base in Central Europe,
> Asia or Africa, but [can] also force the United States,
> for the first time in its history, to a position where nei-
> ther escape nor striking back is possible.

The Chinese also appeared to believe that this power differential was
permanent, given their belief that the Soviet economy was growing
more quickly than economies in the West. For instance, another Chinese journal noted:[61]

> The U.S. may achieve fruitful results in future experi-
> ments and even come to possess both earth satellites
> and intercontinental ballistic missiles. But the Soviet
> Union is advancing at a faster speed than that of the
> capitalist countries. The U.S. is definitely lagging
> behind, and permanently so.[1]

---

[1] The Chinese belief in the inevitable decline of imperialism was also based on the
perception that Western armies and power had suffered a number of setbacks since the
end of World War II. In late 1957, the publication *Jen-min jih-pao (People's Daily)* noted
that "the superiority in strength of socialism over imperialism has been demonstrated
before now in a series of facts. These are: in the Second World War, the main power which
destroyed Hitler and triumphed over the Japanese aggressors was the Soviet Union and
not the combined forces of the United States and Britain. In the Chinese people's war of
liberation, the victor was not Chiang Kai-shek who had the strong support of the United
States, but the revolutionary people of China. In the Korean war, the Chinese People's
Volunteers and the Korean People's Army threw the so-called UN forces . . . back from
the Yalu river. In Vietnam, the Vietnamese Democratic Republic thoroughly defeated the
armed forces of the US-supported French colonialists. In Egypt's struggle to defend its
sovereign rights over the Suez Canal, the Soviet Union's warning to Britain, France and
Israel, coupled with the opposition of world public opinion, played a decisive role in halt-
ing aggression . . . In addition, the decline of the imperialist forces has also been strikingly
manifested in the withdrawal of Britain from India, Burma, Egypt and other colonies, the
withdrawal of the Netherlands from Indonesia, the withdrawal of France from a whole
series of colonies in Western Asia and North Africa. It goes without saying that these with-
drawals resulted from the double blows to imperialism dealt by the socialist forces and the
nationalist forces which oppose colonialism. The superiority of the anti-imperialist forces
over the imperialist forces demonstrated by these events has expressed itself in even more
concentrated form and reached unprecedented heights with the Soviet Union's launching
of the artificial satellites . . . That is why we say that this is a new turning point in the international situation."[62]

The Soviet assessment of the global balance of power in the late 1950s was far more cautious than that of the Chinese. While the Soviets did believe their technological and military achievements signified that the West was now deterred from attacking the USSR, and they repeatedly asserted that the "correlation of forces" was shifting in their favor, under Khrushchev they did not assert that the socialist bloc was militarily more powerful than the West.[63] Additionally, Khrushchev was still extremely impressed with the economic and military potential of the United States, and in speeches he often stressed that the Soviet Union would suffer significant damage in a nuclear war with the West (such references to the potential costs of a nuclear war were typically absent in Chinese leaders' speeches).[64]

The Soviets at this time also became extremely concerned with the possibility of a local war expanding into a global war involving the superpowers, and indeed they began to emphasize a parliamentary path to socialism and a relaxation of tensions with the West.[65] This approach actually preceded the technological developments of 1957; during the Twentieth CPSU Congress the previous year, in addition to denouncing Stalin, Soviet leaders promoted the "three peacefuls:" peaceful competition, peaceful coexistence, and the peaceful transition from bourgeois parliamentary democracy to socialism.[66] From the Soviet perspective, the Chinese were promoting an offensive strategy that was not only dangerous but also premature given their misreading of the global balance of power. Instead, the Soviets believed that nationalist movements and governments in the colonial world would eventually gravitate to the socialist bloc once the Soviet Union (inevitably) overtook the West economically through peaceful competition.[67]

However, from the Chinese perspective, the gradualist approach the Soviets adopted not only represented an abdication of their responsibility to lead "proletarian internationalism," but the refusal to press the advantage against imperialism during its decline jeopardized the final victory of socialism by keeping open the prospect of an imperialist restoration.[68, m] Hence, the Chinese scoffed at the Soviet characterization

---

[m] Another Chinese concern with the Soviet gradualist approach was the fear that the Soviets would use the gains achieved against "imperialism" by national liberation movements and other actors as bargaining chips to reach peaceful coexistence with the United States. Quoting comments made in 1965 by First Premier Zhou Enlai to Enver Hoxha, the leader of Albania, Brazinksy noted that "Zhou launched into a lengthy criticism of Soviet policy: 'A characteristic of modern revisionists . . . is that they are afraid of American

of US President Dwight Eisenhower as a "man of peace"[70] and instead emphasized the need for vigilance in the developing world:[71]

> Can the exploited and oppressed people in the imperialist countries "relax"? Can the people of all the colonies and semi-colonies still under imperialist oppression "relax"? Has the armed intervention led by the U.S. imperialists in Asia, Africa and Latin America become "tranquil"? Is there "tranquility" in our Taiwan Straits when the U.S. imperialists are still occupying our country Taiwan? Is there "tranquility" on the African continent when the people of Algeria and many other parts of Africa are subjected to armed repression by the French, British and other imperialists? Is there "tranquility" in Latin America when the U.S. imperialists are trying to wreck the people's revolution in Cuba by means of bombing, assassination and subversion?

Yet the Chinese critique was not limited to Soviet policies in the colonial world. In particular, the Chinese believed that *they*, and not the Soviet Union, were better suited to lead the struggle against imperialism in the developing world.[72] One of the bases for this belief was writing by Mao during late 1939 and early 1940 in which he spoke of a "new democratic revolution" developing in China and "in all colonial and semi-colonial countries," led (in each country or territory) by the joint dictatorship of several revolutionary classes. These revolutions were different from those that had occurred in Western countries because they were led not by the bourgeoisie but by a popular front movement including all revolutionary groups, and they differed from the Bolshevik revolution because its key enemy was foreign imperialism rather than domestic capitalism.[73] For Mao, the revolutionary alliance leading the revolution would include Communists, nationalists and other radical groups, but with Communists taking the leading role.[74]

---

imperialism and a world war. They are afraid that some local war might escalate, with American interference, into a large scale world war. They do not want the peoples of the world to wage an armed war for their national independence. They are afraid of the peoples of the world revolution. Hence they are trying to discourage and stop such revolutions.' The premier also accused the 'Soviet revisionists' of 'trying to bring the socialist countries, the sister parties, and the national liberation struggles under their control and use them to make compromises with the USA.' "[69]

In contrast, the Soviets argued that the national bourgeoisie must take the leading role in the initial bourgeoisie democratic phase of the revolution, because underdeveloped countries cannot bypass the capitalist stage of development and the bourgeoisie can be useful in destroying "medieval remnants," such as the landlord class, within Asian and African societies. For the Chinese, however, their disastrous experience with the alliance with the Kuomintang in the 1920s taught them that nationalist parties might seek the destruction of their Communist partners and bring about a premature termination of the revolution.[75]

The Chinese also believed that revolutions in the colonial world would resemble the Chinese revolutionary experience rather than the Soviet experience. Zagoria noted:[76]

> The Russian Bolsheviks came to power almost overnight; they took over the major cities and only then expanded their power base to the countryside; inheriting a modern national army from the old regime, they received little experience in guerrilla warfare; they came to give legal revolutionary methods a premium, believing that revolutionary "opportunities" would arise during a time of national crisis, perhaps induced by war, in which the ruling classes would have become so weakened that they would topple almost of their own weight; they had less experience in dealing with the "national bourgeoisie" and more in overthrowing a landed aristocracy.

> The Chinese, by contrast, came to power after a struggle lasting more than two decades; they established bases in rural areas and only then encircled and strangled the cities; their revolutionary experience was almost entirely based on protracted guerrilla warfare; they came to believe that the way to take power was through arduous armed struggle over an extended period of time, a struggle in the course of which the army could be demoralized and the peasantry gradually won over to the side of the revolution; they gave a great deal of attention to the correct handling of the "national bourgeoisie."

The Chinese experience was eventually formalized in the writings of Lin Biao in his discussion of "peoples' wars," which proposed the capture of the countryside before a revolutionary assault on cities.[77] Yet even immediately after the defeat of the Kuomintang in 1949, Chinese leaders suggested that the Chinese revolutionary experience was a model for other movements to follow. In November 1949 Liu Shao-chi, China's head of state and second in command to Mao, noted:[78]

> The course followed by the Chinese people in defeating imperialism and its lackeys and in founding the People's Republic of China is the course that should be followed by the peoples of the various colonial and semi-colonial countries in their fight for national independence and people's democracy. . . . If the people of a colonial or semi-colonial country have no arms to defend themselves they have nothing at all. The existence and development of proletarian organizations and the existence and development of a national united front are closely linked to the existence and development of such an armed struggle. For many colonial and semi-colonial peoples, this is the only way in their struggle for independence and liberation.[n]

China's faith in its qualifications to lead the revolution in the colonial world was also based on a feeling of affinity with Africans and other colonial subjects who shared with China a bitter and humiliating

---

[n] In fact, CCP leaders believed as far back as the 1920s that China was at the forefront of a revolutionary movement against imperialism. Brazinksy noted that "[a]s the CCP grew, so too did its sense of consanguinity with other revolutionaries and its conviction that China was becoming a leader in the global struggle against imperialism. In its weekly journal, the *Guide (Xiangdao)*, the CCP published an article on 5 June calling on all people to 'resist the great massacre of the cruel and savage imperialists.' It argued that the Shanghai massacre was not an 'unusual occurrence;' it was an 'inevitable phenomenon under capitalist imperialist rule.' The party asked: 'Are the small, weak peoples of India, Egypt and Africa as well as the oppressed classes of Europe and the United States not slaughtered regularly by the capitalist-imperialist robbers?' Party leaders believed that the protests sweeping through China would bolster revolutionary forces worldwide. In October 1925, with demonstrations still occurring in major cities, the CCP passed a resolution confidently proclaiming that 'the struggle of the Chinese people has opened a new front against imperialism and at the same time it has increased the strength of the global proletariat and the oppressed peoples of the East.' "[79] The disturbances referred to in the quote are in reference to the 1925 May Thirtieth Movement, which began as a mass protest following the death of a striking Chinese worker by a Japanese foreman in a Japanese mill.

experience with European imperialism and exploitation. For instance, a February 1960 article from the Tsinghua News Agency noted:[80]

> China and Africa have both been subjected for a long time to imperialist plunder and oppression. . . . The peoples of China and Africa are waging a common struggle on two fronts against the same enemy—imperialism. Victories won by either are a support and encouragement to the other.

Additionally, in a speech addressing the second Afro-Asian People's Solidarity Organization (AAPSO) conference in Conakry, Guinea, in April 1960, a Chinese official noted:[81]

> We pay special homage to the heroic Algerian People and their gallant National Liberation Army. They are standing at the forefront of the struggle against imperialism. They have won the admiration and support from the Afro-Asian peoples and the whole world. . . . The imperialists dream they can crush the struggle of the Algerian people by "superior weapons" and have their colonial empires by means of armed suppression. But they are digging their own graves for history has shown that justice always prevails over injustice, the weak over the strong and the newborn force over the decaying one. . . . The Chinese people entertain especially close warm feelings for the African people. . . . We were regarded by the imperialist aggressors as a so-called "inferior race" and our people suffered the same bitterness of slaughter, plundering and enslavement at the hands of foreign colonialists.

Hence, strategic (i.e., differing assessments of the global correlation of power) and tactical differences (i.e., the role of communists in leading the revolution) with the Soviets, combined with a belief in the suitability of their own revolutionary experience for the colonial world and an affinity with colonial subjects, led the Chinese to believe that they, rather than the Soviets, were uniquely qualified to lead the revolution in the colonial world. This belief contributed to a rupture with the Soviets, with the Chinese viewing the struggle in the developing world as a three-way contest between the forces of "imperialism;" "revisionism,"[o] "capitulationism" and "defeatism;" and the genuine forces of proletarian internationalism.[83] Additionally, Mao updated his "two camps" theory by introducing the "Three Worlds Theory," in which the first world consisted of the United States and the Soviet Union; the second world comprised other developed Western countries, primarily in Europe; and the third world consisted of the developing countries in Asia, Africa, and Latin America.[84] China believed itself to be the natural leader of this third grouping.[p]

---

[o] Mao often worried that China would become "revisionist," a concern reflected in this 1963 exchange with a revolutionary from Rhodesia: "African visitor: The Soviets used to help us, and then the red star went out and they don't help us anymore. What I worry about is: Will the red star over Tiananmen Square in China go out? If that happens we will be alone. Mao Zedong: I understand your question. It is that the USSR has turned revisionist and has betrayed the revolution. Can I guarantee to you that China won't betray the revolution? Right now I can't give you that guarantee. We are searching very hard to find the way to keep China from becoming corrupt, bureaucratic and revisionist. We are afraid that we will stop being a revolutionary country and will become a revisionist one. When that happens in a socialist country, they become worse than a capitalist country. A Communist party can turn into a fascist party. We've seen that happen in the Soviet Union. We understand the seriousness of this problem, but we don't know yet how to handle it."[82]

[p] Other scholars have taken a more conventional realist analysis of Chinese motivations by emphasizing the inevitability of the jockeying for power between two large countries that sought enhanced global status irrespective of their domestic political, economic, and social orientations. Chau noted that "sinologists often cite the ideological aspects of China's foreign policy when referring to the Maoist period, including in Africa. Mao himself was repeatedly perceived as a radical Communist firebrand . . . However, the primary rationale for the Chinese-Soviet riff was not ideological but strategic. Located geographically on the same Eurasian landmass, the two historically proud and geographically large nations, with similar aspirations of becoming great powers, were bound to clash."[85] Taylor noted that "the PRC has been called 'the high church of realpolitik.' China's sense of inadequacy on a global scale vis-à-vis other major powers and its aspirations to be ranked as a true 'great power' has therefore informed Beijing's policy towards the developing world. By employing rhetoric, the development of political and commercial linkages and a general sense of identification with the developing world, the People's Republic of China has sought to boost its own international standing and maximize its options in response to the changing dynamism of the international system."[86] Regarding the "high church of realpolitik," see Christensen.[87] Hence, China's efforts to rally the developing world, rather than representing a moral crusade against "imperialism" and "revisionism," can instead be seen as an effort to boost its global influence vis-à-vis the Soviet Union and the United States. Additionally, it comes as no surprise that during this period China was always fearful that the Soviets would bargain away their interests in negotiations with the United States, because any agreement reached between the two superpowers may have limited China's ability to acquire power and influence. Hence, China was critical of the July 1963 signing of the nuclear test ban treaty among the United States, the Soviet Union, and the United Kingdom. The Chinese, who first tested a nuclear weapon in October 1964, labeled the

RESISTANCE MOVEMENT SELECTION

Which were the main factors influencing the decisions by Moscow and Beijing to support specific liberation movements in Southern Africa? In the case of the Soviet Union, it helps to first locate national liberation movements within the Soviet conception of the world revolutionary movement. As Kempton noted, for the Soviets, such movements composed the global national liberation movement and formed one component of the world revolutionary process.[90] The other two components consisted of the Socialist community of nations led by the Soviet Union and the international worker's and Communist movement, consisting of Marxist-Leninist parties and led by the CPSU.[91] The Soviets also distinguished between groups within the global national liberation movement. The first group included "revolutionary democratic" parties, such as the Movimento Popular de Libertação de Angola (or the People's Movement for the Liberation of Angola, MPLA) and the Frente de Libertação de Moçambique (Mozambique Liberation Front, or FRELIMO), that differed from conventional Communist parties because their membership reflected a more diverse class background and since such groups often lacked a cohesive doctrine and discipline.[92] The second group included movements focused primarily on national liberation, such as the ANC of South Africa, SWAPO of Namibia, and the Palestine Liberation Organization. The last category consisted of ruling or nonruling Socialist-leaning parties, such as ZANU (after it attained power) and the Chilean Socialist Party.[93]

---

treaty "a big fraud to fool the people of the world," and it represented an attempt by the signatories to "consolidate their nuclear monopoly and bind the hands of all peace-loving countries subjugated to the nuclear threat," as well as a plot by Western "imperialists" and Soviet "modern revisionists" to dominate the world.[88] The PRC's initial forays into Africa commenced in 1955 with the Afro-Asian Conference in Bandung, Indonesia, which featured Chinese representation and representatives from six African states, and Sino-African interactions intensified in the early 1960s and culminated with Foreign Minister Zhou Enlai's two-month tour of Africa from December 1963 to February 1964. Alden and Alves noted that the establishment of a number of new independent states in Africa presented an opportunity for China to solve what they termed China's "legitimacy problems," because China at this time was not recognized by the United Nations (UN) or the United States, and indeed it was the target of efforts by the United States and later by the Soviet Union to isolate it internationally. Fourteen African countries established diplomatic relations with the PRC in the 1960s and twenty-two did so in the 1970s, and the African vote played a prominent role in the 1971 vote in the UN that led to China taking over Taiwan's permanent seat on the UN Security Council. Twenty-six African states voted for the PRC's entry into the UN, and in fact seven African states changed their "no" vote to a "yes" vote.[89]

In reviewing past support for various opposition groups, Kempton noted that the Soviets tended to pursue either a "model" or an "ally" strategy. In the former, the Soviets supported groups that accepted the Soviet political and economic model, including the need for a vanguard party to lead society toward communism, and this strategy came in two variants.[94] The revolutionary variant entailed the USSR working with an existing Communist party or Marxist revolutionary group before it comes to power, whereas the nonrevolutionary variant called on the USSR to encourage a movement to adopt the Soviet model. One of the main benefits of this strategy is building influence, because the close collaboration between the USSR and a revolutionary group, to include the USSR's provision of significant financial, military, and human resources, provided the Soviets with various entryways to build influence with a group. This influence could translate into significant leverage in a country as a whole were the group to assume power.

Additionally, given widespread political instability in the developing world at the time, a model strategy was a form of risk reduction because it ensured that the movement and the country (if the group assumed power) remained in the Soviet orbit after any leadership changes. By developing close ties with leading members of a movement below the senior leadership level, to include Soviet military and ideological training in the USSR, Eastern Europe, or Cuba, the Soviet Union established close ties with the next generation of leaders.[95] Furthermore, the adoption of the Soviet political and economic model might lead to its institutionalization within the movement and the country itself, thereby reducing Soviet dependence on the survival of any particular leader.[96] Such concerns were not trivial; in the late 1950s the Soviets developed ties to a number of radical leaders, including Modibo Keita of Mali, Patrice Lumumba of the Congo, Ahmed Ben Bella of Algeria, and Gamal Abdel Nasser of Egypt, whose replacements led to sharp declines in Soviet influence.[97] Indeed, Anwar Sadat, who succeeded Nasser in 1970, expelled thousands of Soviet advisors two years later.

Although a model strategy can potentially reduce political risk, it has various drawbacks. First, historically it has been very costly. Soviet support to countries such as Cuba, Afghanistan, and Vietnam amounted to billions of dollars annually in each case.[98] Additionally, some countries that have modeled themselves after the Soviet Union, such as China, Albania, and Romania, established a level of independence from the USSR despite their Communist orientations.[99] Hence, at times the Soviets adopted what Kempton called an ally strategy, in which a movement or government was deemed worthy of Soviet support based on its foreign policy orientation, specifically whether it was pro-Soviet and, ideally, anti-American, irrespective of the arrangement of its political and economic systems.[100] In addition to being less costly, the adoption of an ally strategy was at times an acknowledgment that in many cases Communist parties in the Third World lacked sufficient legitimacy and support to attain and maintain power.[101] Additionally, the use of a model strategy, such as in Angola and Afghanistan, led to increasing tensions with the West.[102]

In practice, both the model and ally strategies represent ideal types, with most Soviet relationships with liberation movements falling somewhere between the two. Yet greater emphasis has been placed on one or the other strategy at different times throughout Soviet history. Kempton noted that Lenin essentially advocated an ally strategy by arguing for a two-stage revolutionary process.[103] Given the relative weakness of Communist parties in the developing world, Lenin argued that in the first stage, Communist parties should cede leadership of the revolution to the national bourgeoisie, whose interest in independence was shared by the USSR and who sought the weakening of European colonial powers. Yet this revolutionary alliance would fracture in Lenin's second stage (occurring at some undetermined time in the future), which required Communist parties to take the lead of the liberation movement once independence was achieved and begin the Socialist transformation of society. In the meantime, before the second stage, Lenin advocated that Communist parties criticize rightist elements within the liberation movement, and he indicated that bourgeoisie-led movements were worthy of Soviet support as long as the movements were genuinely revolutionary and did not obstruct efforts by Communists to educate and mobilize the populace.[104]

The Soviet use of an ally strategy at this time was limited to its relations with China and Germany, and the two-stage thesis was accepted until the nationalist Kuomintang massacred Chinese Communists in April 1927. After this disaster, during the Sixth Congress of Comintern in 1928, Stalin adopted the revolutionary alternative of the model strategy by limiting aid to Communist parties loyal to the USSR.[105] Yet Soviet enthusiasm for the model strategy waned with the rise of fascism in Europe and the Japanese invasion of Manchuria in the 1930s. The need for an alliance between revolutionary and liberal democracy against fascism led the Soviet Union to call on Communist parties in the developing world to support the war effort. It also led to the ending of rhetorical support for revolutions in the developing world (because such movements would hurt the interests of France and England, who were allies of the Soviet Union during World War II).[106] Indicative of this new line was Soviet support for a united front of the Kuomintang and the CCP to fight the Japanese occupation, and aid was limited to those groups regarded as critical for terminating the war.[107]

After World War II, the Chinese Communists' victory over the Kuomintang and the prominence of the two-camps thesis in which the world was divided into one camp characterized by "peace, socialism and democracy" and the other by "capitalism, imperialism and war" led the Soviets to return to the model strategy.[108] This shift explains the scorn and ridicule heaped toward non-Communist nationalist leaders during the postwar Stalinist era. Yet the rise of Khrushchev brought an important shift toward an ally strategy, as Khrushchev saw anti-Western leaders in the developing world as potential allies who over time could be encouraged to adopt the Soviet political and economic model.[109] Hence, the doctrine of "different roads to socialism" was promulgated during Khrushchev's rule.[110] Consistent with this new thrust, in the summer of 1961 Khrushchev and the Central Committee of the CPSU adopted a new and aggressive KGB strategy calling for the active use of national liberation movements in the developing world to further Soviet interests in the struggle with the West.[111, q]

It is within this context that Soviet links with ZAPU (more specifically, its predecessor) were established. The first links appear to have occurred in late 1959 or early 1960, when Nkomo first requested aid from the Soviet embassy in London.[113]

In the case of China, its involvement in the sponsorship of African resistance movements coincided with the Sino-Soviet rift, and thus while Taylor noted that China did not rule out aiding ZAPU in the early 1960s, the sponsorship of ZANU was perhaps a better fit because ZANU was not tainted by Soviet influence.[114, r] Indeed, the timing was

---

q This new KGB strategy appears to have involved Rhodesia. Shubin noted that in July 1961 the then-KGB chairman Alexander Shelepin proposed to Khrushchev the provision of training and arms to African rebels, including those in Rhodesia, to foster anti-colonial uprisings.[112]

r Nonetheless, there were some interactions between China and ZAPU. In 1964 James Chikerema, then one of Nkomo's key subordinates, visited China, and at that time perfect, with one author noting that "China's search for clients coincided with ZANU's search for patrons"[116] and another noting that the sponsorship of ZANU represented an opportunity for Beijing to promote anti-Sovietism in Africa.[117]

Overall, though, it must be recognized that China's support to African liberation movements was largely rhetorical and symbolic. It lacked the financial resources to fully support ZANU and other African liberation movements, and the weapons it provided were seen as inferior relative to Soviet military aid.[118,s] Yet for Africans the PRC served as a model for what can be achieved by a backward agrarian economy in a relatively short period. Although China's economic developments in the 1950s and 1960s pale in comparison to the extraordinary growth it has achieved since Deng Xiaoping liberalized its economy in the late 1970s, by the 1960s Chinese industrial (and nuclear) achievements sufficiently impressed African elites, who were acutely aware that the level of Chinese economic development and economic and social conditions within China were similar to those found in sub-Saharan Africa. In 1965 Julius Nyerere, president of Tanzania, noted:[120]

> The vast majority in both China and Tanzania earn their living from the land or in the rural areas. And both of us have only recently won freedom from that combination of exploitation and neglect which characterizes feudal and colonial societies. We have therefore much to learn from each other.[t]

---

some ZAPU troops were receiving training in Nanking, China. Additionally, some ZAPU troops received training in Tanzanian camps staffed by Chinese instructors. In contrast, ZANU was largely shunned by the Soviet Union. Soviet officials did meet Herbert Chitepo, the ZANU national chairman, on the sidelines of the April 1973 International Conference on Southern Africa in Oslo, Norway, but overall the Soviet stance was that ZANU would have to renounce its ties with China if it wanted to receive Soviet aid and, in particular, Soviet weapons.[115]

  [s] As discussed in greater depth in chapter 4, the poor quality of Chinese arms relative to Soviet military aid was one of the leading causes of a 1974 factional feud in ZANU known as the Nhari rebellion. The inability of ZANU to directly obtain military aid from the Soviet Union led regional leaders such as Julius Nyerere of Tanzania and Samora Machel of Mozambique to appeal directly to the USSR to provide military assistance to ZANU. These appeals were unsuccessful, yet nonetheless ZANU was able to obtain some Soviet weapons from Mozambique and Ethiopia without Soviet consent.[119]

  [t] Additionally, after an October 1959 visit to Beijing, the Guinean minister of education noted that he was "quite convinced of the efficacy of Chinese methods. I was greatly impressed by the similarity of the economic problems that China has succeeded in solving

NOTES

[1] Christopher Andrew and Vasili Mitrokhin, *The World Was Going Our Way: The KGB and the Battle for the Third World* (New York: Basic Books, 2005), 423.

[2] Edward T. Wilson, *Russia and Black Africa Before World War II* (New York and London: Holmes & Meier Publishers, 1974), 94.

[3] Ibid., 95.

[4] Quoted in Surendra K. Gupta, *Stalin's Policy Towards India, 1946-1953* (New Delhi: South Asian Publishers, 1988), 4.

[5] James Bunyan and H.H. Fisher, ed., *Bolshevik Revolution, 1917-1918; Documents and Materials* (Stanford: Stanford University Press; H. Milford, Oxford University Press, 1934), 243–244.

[6] Adam B. Ulam. *Expansion and Coexistence: Soviet Foreign Policy 1917-1973*, 2nd Edition (New York: Holt, Rinehart and Winston, Inc., 1974), 54–55.

[7] Jon Jacobson, *When the Soviet Union Entered World Politics* (Berkely, CA: University of California Press, 1994), 13.

[8] Ulam, *Expansion and Coexistence*, 13.

[9] Ibid., 27.

[10] Wilson, *Russia and Black Africa Before World War II*, 94–95.

[11] Daniel R. Kempton, *Soviet Strategy toward Southern Africa: The National Liberation Movement Connection* (New York: Praeger, 1989), 19.

[12] Wilson, *Russia and Black Africa Before World War II*, 122.

[13] Ibid.

[14] Robert F. Gorman, "Soviet Perspectives on the Prospects for Socialist Development in Africa," *African Affairs* 83, no. 331 (1984): 168.

[15] V. I. Lenin, *Collected Works*, vol. 30, 159, quoted in Gorman, "Soviet Perspectives on the Prospects for Socialist Development in Africa," 168.

[16] Ibid.

[17] Wilson, *Russia and Black Africa Before World War II*, 126.

[18] Ibid., 128.

[19] Ibid., 129.

[20] Ibid., 104, 110–111.

[21] Ibid.

[22] Ibid., 130.

[23] Ibid.

[24] Ibid., 130, 136.

[25] Gorman, "Soviet Perspectives," 167.

[26] Andrew and Mitrokhin, *The World Was Going Our Way*, 423–424

[27] Gorman, "Soviet Perspectives," 166–167.

and those that are now facing the peoples of Africa. In China, I saw what can be done if you mobilize the vital forces of a nation. With all due regard to the difference in magnitude we now propose to do the same thing."[121]

28 Andrew and Mitrokhin, *The World Was Going Our Way*, 423; and Vladimir Shubin, *The Hot "Cold War": The USSR in Southern Africa* (London: Pluto Press, 2008), 239.

29 Gorman, "Soviet Perspectives," 170.

30 Amal Jayawardena, "Soviet Involvement in South Asia: The Security Dilemma," in *Security Dilemma of a Small State: Sri Lanka in the South Asian Context*, ed. P. V. J. Jayasekera (New Delhi: South Asian Publishers Pvt. Ltd., 1992), 374.

31 Gorman, "Soviet Perspectives," 170.

32 Andrew and Mitrokhin, *The World Was Going Our Way*, 426.

33 Gorman, "Soviet Perspectives," 170.

34 Ibid., 173.

35 Ibid., 171.

36 Kempton, *Soviet Strategy*, 19.

37 Andrew and Mitrokhin, *The World Was Going Our Way*, 427.

38 Sue Onslow, "The Cold War in Southern Africa: White Power, Black Nationalism and External Intervention," in *Cold War in Southern Africa: White Power, Black Liberation, ed. Sue Onslow* (London and New York: Routledge, 2009), 19–20.

39 Ibid., 20.

40 David E. Albright, "The Communist States and Southern Africa," in *International Politics and Southern Africa*, ed. Gwendolen M. Carter and Patrick O'Meara (Bloomington, IN: Indiana University Press, 1982), 17.

41 Albright, "The Communist States and Southern Africa," 17–18.

42 *The Role of the Soviet Union, Cuba, and East Germany in Fomenting Terrorism in Southern Africa: Hearings Before the Subcommittee on Security and Terrorism of the Committee on the Judiciary*, United States Senate, 97th Cong., 2d sess. (1982), 617.

43 N. Friedman, *The Fifty-Year War: Conflict and Strategy in the Cold War* (Annapolis: Naval Institute Press, 2000), 399.

44 Albright, "The Communist States and Southern Africa," 18.

45 Chris Alden and Ana Cristina Alves, "History & Identity in the Construction of China's Africa Policy," *Review of African Political Economy* 35, no. 115 (2008): 46.

46 Chen Jian, *Mao's China and the Cold War* (Chapel Hill: University of North Carolina Press, 2001), 7, 11–12, 47.

47 Gregg Brazinsky, *Winning the Third World: Sino-American Rivalry During the Cold War* (Chapel Hill: University of North Carolina Press, 2017), 16.

48 Brazinsky, *Winning the Third World*, 17; Michael H. Hunt, *The Genesis of Chinese Communist Foreign Policy* (New York: Columbia University Press, 1996), 114.

49 Ibid., 44.

50 Alden and Alves, "History and Identity," 45, 48.

51 Jian, *Mao's China*, 50.

52 Quoted in Jian, *Mao's China*, 50.

53 Quoted in Alaba Ogunsanwo, *China's Policy in Africa, 1958–71* (London: Cambridge University Press, 1974), 15.

54 Jian, *Mao's China*, 26–28, 65.

55 Ibid., 64–66.

[56] Ibid., 9, 68; and Donald Zagoria, *The Sino-Soviet Conflict, 1956–1961* (Princeton, NJ: Princeton University Press, 1962), 16.

[57] Zagoria, *Sino-Soviet Conflict*, 53, 63.

[58] Ibid., 63,70.

[59] Ibid., 158–159, 167.

[60] Quoted in Zagoria, *Sino-Soviet Conflict*, 162–163.

[61] Ibid., 163.

[62] Ibid., 164–165.

[63] Zagoria, *Sino-Soviet Conflict*, 157, 159.

[64] Ibid., 159.

[65] Ibid., 168, 251.

[66] O. Igho Natufe, *Soviet Policy in Africa: From Lenin to Brezhnev* (Bloomington, IN: iUniverse, Inc., 2011), 153.

[67] Zagoria, *Sino-Soviet Conflict*, 153, 258.

[68] Ibid., 153, 257.

[69] Brazinsky, *Winning the Third World*, 233.

[70] Ogunsanwo, *China's Policy in Africa*, 64, 69.

[71] Ibid., 66–67.

[72] Zagoria, *Sino-Soviet Conflict*, 267.

[73] Ibid., 248–249.

[74] Ibid., 249–250.

[75] Ibid., 247, 253.

[76] Ibid., 16–17.

[77] Quoted in Ian Taylor, *China and Africa: Engagement and Compromise* (London: Routledge, 2006), 107.

[78] Quoted in Ogunsanwo, *China's Policy in Africa*, 18–19.

[79] Brazinsky, *Winning the Third World*, 20.

[80] Ibid., 65.

[81] Ogunsanwo, *China's Policy in Africa*, 98.

[82] Quoted in David H. Shinn and Joshua Eisenman, *China and Africa: A Century of Engagement* (Philadelphia: University of Pennsylvania Press, 2012), 62–63.

[83] Ogunsanwo, *China's Policy in Africa*, 113.

[84] Alden and Alves, "History and Identity," 48.

[85] Donovan C. Chau, *Exploiting Africa: The Influence of Maoist China in Algeria, Ghana and Tanzania* (Annapolis: Naval Institute Press, 2014), 19.

[86] Taylor, *China and Africa*, 6.

[87] Thomas J. Christensen, "Chinese Realpolitik: Reading Beijing's World-View," *Foreign Affairs* 75, no. 5 (September–October 1996): 37.

[88] Natufe, *Soviet Policy in Africa*, 159.

[89] Alden and Alves, "History and Identity," 47, 51; and George T. Yu, "China's Role in Africa," *Annals of the American Academy of Political and Social Science* 432, no. 1 (1977): 105.

[90] Kempton, *Soviet Strategy*, 4.

91 Ibid.

92 Ibid.

93 Ibid., 4–5.

94 Ibid., 11.

95 Ibid., 13.

96 Ibid., 12.

97 Ibid., 11.

98 Ibid., 15.

99 Ibid., 14.

100 Ibid., 17.

101 Ibid.

102 Ibid.

103 Ibid., 5, 19–20.

104 Ibid., 20.

105 Ibid., 20–21.

106 Ibid., 22.

107 Ibid.

108 Ibid.

109 Ibid., 23.

110 Ibid.

111 Andrew and Mitrokhin, *The World Was Going Our Way*, 432.

112 Vladimir Shubin, "Moscow and Zimbabwe's Liberation," *Journal of Southern African Studies*, 43, no. 1, (2017), 226.

113 Kempton, *Soviet Strategy*, 97.

114 Ian Taylor, *China and Africa: Engagement and Compromise* (London: Routledge, 2006), 107.

115 Kempton, *Soviet Strategy*, 102, 126; and Keith Somerville, "The U.S.S.R. and Southern Africa Since 1976," *Journal of Modern African Studies* 22, no. 1 (1984): 93–94.

116 Sabelo J. Ndlovu-Gatsheni, "Angola–Zimbabwe Relations: A Study in the Search for Regional Alliances," *The Round Table* 99, no. 411 (2010): 637–638.

117 Taylor, *China and Africa*, 107.

118 Yu, "China's Role in Africa," 101.

119 Kempton, *Soviet Strategy*, 126.

120 Quoted in Yu, "China's Role in Africa," 101.

121 Quoted in Ogunsanwo, *China's Policy in Africa, 1958–71*, 31.

CHAPTER 3.
HISTORICAL CONTEXT

THE PHYSICAL LANDSCAPE

Terrain

Most of Africa sits on a vast plateau that runs from the Cape in the south to the Atlas Mountains in the north, with a narrow coast belt merging the continent with the oceans.[1] It is immensely rich in minerals, spurring high interest in the continent from a parade of external economic and political actors. Before the industrial revolution, southern Africa's strategic vantage point overlooking one of the great sea lanes of the maritime trade routes between Europe and the Far East lured British, Dutch, and Portuguese explorers and traders.

Figure 3-1. Map of area surrounding Rhodesia.

Rhodesia occupied a small section of the vast African plateau. Interestingly, its borders were not created as a result of the Berlin Conference of 1884, during which the European powers with colonial interests in Africa carved up the continent. Rhodesia's borders were created by the natural boundaries that confine its territory (see Figure 3-2). To the north is the Zambezi River separating Rhodesia from its once partner colony, Northern Rhodesia (now Zambia). With the exception of two bridges at Victoria Falls and Chirundu and the dam wall at Kariba Dam, the only way to cross the Zambezi is by boat. Demarcating the border with South Africa is the Limpopo River, which is easily traversed during the dry season (April to September) and only slightly more difficult to traverse during the wet season (November to March). On the western border, the Kalahari Desert created a natural border with what was Bechuanaland (now Botswana). Although this border offered fair potential for infiltration into Rhodesia during the insurgency, Botswana's military and border guards regularly patrolled Botswana's Rhodesian border in hopes of deterring cross border activity from either direction. To the east, the Eastern Highlands form the border between Mozambique and Rhodesia.

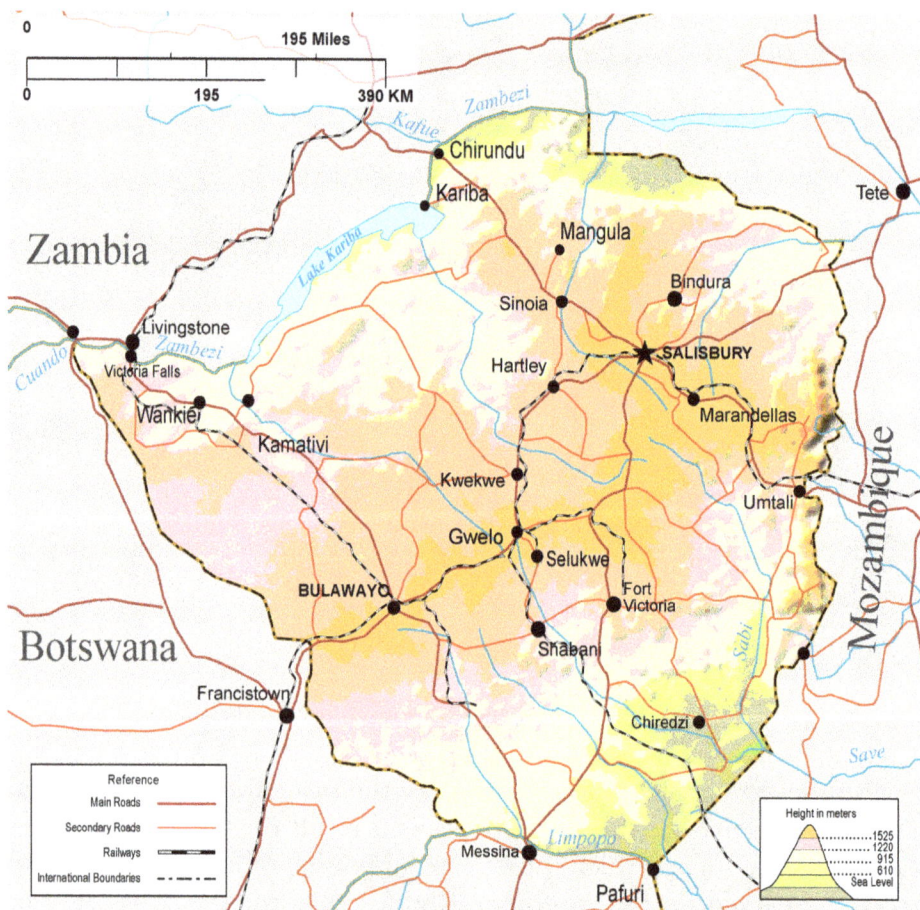

Figure 3-2. Map of Rhodesia.

Climate

Rhodesia's location on the globe would suggest a tropical climate. However, with a quarter of its land mass lying above four thousand feet above sea level and the vast majority above two thousand feet, Rhodesia's climate and topography have an abundance of variety.[2] Fertile rolling plains in the center of the country stand in sharp contrast to the tall formidable escarpments along the Zambezi River banks in the north of the country. Along the eastern border mountains and dense forests dominate. The low veld areas along the Zambezi and the Limpopo in the south also present climatological variations. Throughout the African spring and summer (September to March), temperatures in the high veld central regions may exceed one hundred degrees in the shade, with considerably higher marks set in the low veld along the rivers.[3] Rainfall in Rhodesia is generally confined to the rainy season of November through March, during which time twenty-five to thirty inches of rain may fall, on average. During the insurgency, these weather factors added to the various geographical features to influence the conduct of operations for both the insurgents and the Rhodesian government.

Demographics

Estimates for 1965 indicate a population of approximately 3 million, consisting of a black African population of more than 2.5 million living in reserved tribal lands or townships surrounding white towns, which were populated by approximately 270,000 whites at that time. Additionally, 7,000 white farmers operated commercial farms.[4] The demographic divide continued to increase as the conflict matured. In 1973 the racial breakdown of the population was 5.8 million blacks and 273,000 whites.[5] The urban black population was 814,000, and the white urban population was 213,000.[6] The combination of geography and land distribution created a fairly dispersed rural population. This dynamic resulted in a seminatural segregation: the rural countryside contained a majority of the black population, while the cities contained the majority of the white population.

Socially, the challenge for the nationalist movements was the diverse ethnic makeup of the population. While it is common to divide the black population of colonial Rhodesia between the Ndebele and the Shona, the two major tribal groups within the country, there are several other smaller tribes (the Tonga, the Venda, and the Shangaan), as well as various subtribes within the Ndebele and the Shona.[7] Tribal affiliation in turn impacted loyalties to the nationalist parties.

The histories of the two major tribal groups are also important elements of Rhodesian society throughout the colonial period to the commencement of majority rule in 1980. The Shona made up roughly 80 percent of the black African population and had a centuries-long history in areas of north and east Rhodesia.[8] They were actually a collection of cultural groups with agriculturalist and pastoralist traditions. The Ndebele, on the other hand, made up just fewer than 20 percent and were relative newcomers to the area known as Rhodesia. They arrived in the late 1830s, having broken off from Zulu tribes living in northern and eastern South Africa. They were a highly centralized militaristic society that lived off raiding neighboring peoples.[9] The Ndebele people are credited with identifying the people they raided as "Shona," and the label was adopted by the British explorers who first met with Ndebele Chief Lobengula in Bulawayo in 1888 to negotiate mining concessions in his territory.

The distinctions between these two groups were significant. Although some members of each nationalist party crossed mainstream tribal affiliations (Ndebele in Zimbabwe African National Union, or ZANU; and Shona in Zimbabwe African People's Union, or ZAPU), for most individuals these tribal affiliations carried powerful identities and loyalties. The Shona, for example, place historical blame on the Ndebele for granting concessions to Cecil Rhodes's British South Africa Company, and they harbor historical resentment for Ndebele warriors' frequent attacks on Shona settlements before the arrival of British settlers.

THE ECONOMY

Rhodesia's economy was largely based on cattle farming, agriculture (tobacco and maize), and mining (asbestos, gold, coal, chrome, copper, cobalt, and lithium). Not surprisingly, access to favorable land, whether for farming or mining, played a major role in the conflict. The Land Apportionment Act of 1930 established a division of land as seen in Table 3-1.

Table 3-1. Land distribution 1931.

	Acres	Percentage of Land
White (settler) areas	49,149,000	50.8
Native reserves	21,600,000	22.3
Native purchase areas	7,465,566	7.7
Forest areas	591,000	0.6
Unassigned areas	17,793,300	18.4
Undetermined areas	88,000	0.1
Total	98,686,866	100.0

Source: Martin and Johnson, *The Struggle for Zimbabwe*, 53.

"Native reserves" consisted of territory where black Africans were "temporarily" settled, while "native purchase areas" consisted of territory set aside for potential purchase by black Africans.[10,a] The intent of the Land Apportionment Act was to ensure that cities, towns, and commercial areas were exclusively white areas with few, if any, black Africans, who themselves were confined to sprawling townships whose stark poverty contrasted with the evident wealth of the white areas.[12] By the early 1960s, Europeans, who constituted one-seventeenth of the population, held more than one-third of the land in Southern Rhodesia, with most of the best land held by 6,400 white farmer-owners and 1,400 white tenant farmers. This allocation is depicted in Figure 3-3, which shows that all the cities and the most fertile arable land lie in the white-colored areas of the map, which designate land reserved for Europeans. The orange areas, known as Tribal Trust Lands (TTL), were in areas of generally lower agricultural fertility and were more difficult to access.[13] The native purchase areas in blue were those areas in which Africans could purchase land from the Tribal Trust.

The impact of the Land Apportionment Act and its numerous successors profoundly affected race relations throughout the twentieth century. As Blake points out, although some segregation of races

---

a   Martin and Johnson noted that black farmers had to possess a Master Farmers Certificate to purchase land in the native purchase areas (no such requirement applied to white farmers), and that 4 million of the 7.5 million acres available for purchase were in remote areas of the country and unsuitable for farming.[11]

would have happened naturally because of the cultural, economic, and social proclivities of the respective groups, under the Land Apportionment Act many thousands of Africans were compulsorily moved from land they had occupied for generations.[14] The law became a symbol and the very embodiment of everything most resented in European domination.

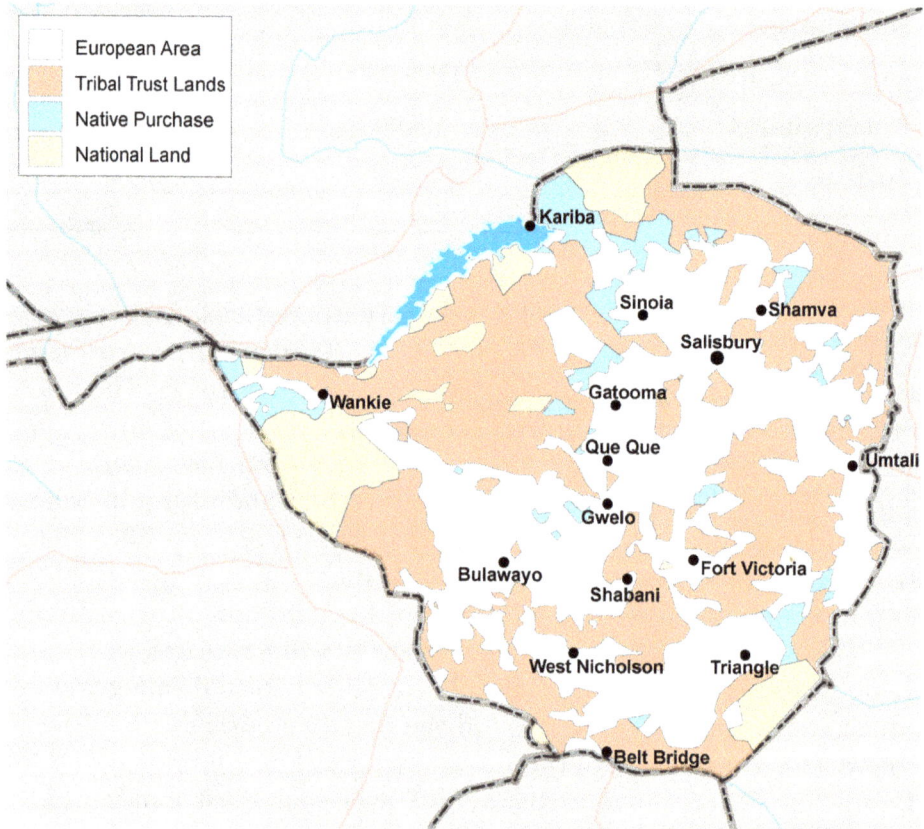

Figure 3-3. Rhodesia land distribution 1965.

THE ROAD TO REBELLION

To begin to understand the insurgency that African nationalists waged against white Rhodesians, it is necessary to understand the history of the white population in Rhodesia and its interaction with the indigenous black population of the country throughout nine decades of white minority rule.

In 1867 gold was discovered in the territory that would become Rhodesia. Several mining interests in South Africa, as well as Portuguese entities in Mozambique, approached King Lobengula, the leader of the Matabele tribe controlling much of the region, to secure mineral rights to the land containing the suspected gold deposits. Through a series of moves and agents, Cecil Rhodes succeeded in winning the prize. The agreement, known as the Rudd Concession, would be the genesis for a series of conflicts that occurred between white settlers and native Africans for much of the 1890s. As detailed by Baxter[15] and Blake,[16] the discrepancies between Rudd's verbal assurances to Lobengula and what was contained in the written concession Lobengula ultimately signed on October 18, 1888, were life altering for the Matabele king and his people, as well as for the Mashona tribe that shared this part of Africa with the Matabele.

Fearful of losing control of his territory, Lobengula received assurances that only ten miners would enter the country to conduct mining operations, that the activities would not be located near villages, and that the miners would abide by the laws of Lobengula's tribe.[17] In exchange for his signature, Lobengula was promised one thousand Martini-Henry rifles with one hundred thousand rounds of ammunition, a gunboat for patrolling the Zambezi (or alternatively £500), and a £100 monthly payment for the rest of his life.[18] On October 29, 1889, Queen Victoria signed the royal charter granting Rhodes's British South Africa Company sole administrative rights to the territory covered by the Rudd Concession.[19] When the time came to implement the terms of the agreement, the original pioneer column that entered Lobengula's territory numbered seven hundred, not ten as promised.

By 1908, the dreams of gold riches had been proven false. Three rebellions of the Matabele and Mashona, in 1893 and 1896, over the increasing presence of white settlers and their acquisition of land made it imperative that the company shift its focus from seeking illusory gold to securing its hold on the territory. Thus, the company prioritized European agricultural settlement as the economic engine of the colony. As a part of the original division of land designed by Leander Starr Jameson, the first administrator of Rhodesia, native reserves were established in an attempt to segregate the natives from the white settlers. Of the ninety-six million total acres that was Rhodesia, 4.1 million were set aside in Matabeleland for native occupation in 1896. This amount was adjusted over the ensuing twenty years until it reached a total of 20.5 million acres in Matabeleland and Mashonaland in 1914.[20]

In 1923, the white settlers gained what would become a pivotal status within the British Empire. While the British South Africa Company was seeking to relieve itself of the burdensome administrative costs of running the colony, the settlers were seeking more control over their political existence. South Africa was lobbying the British government to form a union between South Africa and its northern neighbor Southern Rhodesia. The settlers were asking for the right to govern themselves under a status known as "Responsible Government." After a referendum that rejected union with South Africa, Southern Rhodesia was granted Responsible Government under a new constitution. This designation allowed the white population to assume full control of the administrative, economic, social, and political life of Rhodesia, thereby making it a self-governing British colony. The white settlers, rather than the British Colonial Office (or the black African population), controlled the political system, and this control was granted by the British government.

In 1953, the Federation of Rhodesia and Nyasaland, also known as the Central African Federation (CAF), was established (see Figure 3-4). The federation consisted of the self-governing colony of Southern Rhodesia and the protectorates of Northern Rhodesia and Nyasaland. At the time of its establishment, Britain retained jurisdiction in Northern Rhodesia and Nyasaland over matters that impinged directly on the lives of ordinary Africans.[21]

With pressure to end colonialism from the UN and the Organisation of African Unity (OAU), the British government was increasingly speaking about the reality and inevitability of decolonization. During a speech in Cape Town in 1960, British Prime Minister Harold Macmillan stated, "The wind of change is blowing through the continent. Whether we like it or not, this growth of national consciousness is a political fact. We must accept it as fact. Our national polices must take account of it."[22]

Figure 3-4. Map of Central African Federation.

In 1960 growing discontent among nationalists in all three federation entities led the British government to appoint a commission
to assess the status of the federation. The commission, known as the
Monckton Commission, or more fully, the Advisory Commission on the
Review of the Constitution of Rhodesia and Nyasaland, unequivocally
reported that racial mistrust and hostility were so intense that the CAF,
in its 1960 form, could survive only by using force or introducing fundamental changes in the racial policies of Southern Rhodesia. These
changes would include an expanded franchise, the establishment of
racial parity in the federal house, the granting of self-government to
Nyasaland and Northern Rhodesia, and the barring of any constitutional revisions without full African agreement.[23]

No one in the British government saw force as a reasonable alternative, and the Monckton Commission report categorically ruled it out in its assessment: "To hold the Federation together by force we regarded as out of the question."[24] Macmillan's comments and the Monckton Commission report signaled a clear mood in Britain toward relinquishing control of Nyasaland and Northern Rhodesia to the majority African leaders.

By comparison with its neighboring countries, the nationalist movement was late coming to Southern Rhodesia. Early organizations, such as the City Youth League, were exclusively focused on urban centers with agendas limited to addressing urban and worker grievances such as the bus boycott in Salisbury in 1956.[25,b] They were not focusing on the larger goal of majority rule or independence that so dominated the African movements in other British colonies at the time. The first mass party, the Southern Rhodesia African National Congress (SRANC), formed in 1957.[26] As nationalist movements in Southern Rhodesia gained momentum, their leaders failed to comprehend the constraints of the political environment and overestimated the British government's ability and willingness to support their cause as it had in Northern Rhodesia and Nyasaland.[27]

Concerned that SRANC disturbances in 1959 were connected to similar acts in Nyasaland and Northern Rhodesia, the Rhodesian government declared a state of emergency and banned the SRANC in February 1959.[28,c] After outbreaks of violence in 1960, the nationalists formed the National Democratic Party (NDP). The first elected leader was Joshua Nkomo, former head of SRANC who had escaped the government roundup in 1959 by being out of the country at the time of the SRANC's banning.

The primary goal of the NDP was to enlist British support for majority rule and to be included in the 1961 constitutional conference

---

[b] The Salisbury bus boycott was organized to challenge the bus fare rates paid by African laborers commuting to jobs in the capital.

[c] By 1959 the success of the CAF, which comprised Northern Rhodesia (Zambia), Nyasaland (Malawi), and Southern Rhodesia, was in serious doubt. African nationalist movements in Northern Rhodesia and Nyasaland were becoming increasingly violent in calling for racial equality in politics, economics, and society. These activities were led by the respective branches of the ANC. When the SRANC began to undertake activities in solidarity with their federation brethren, the government of Southern Rhodesia took the measures described above to retain control.

held in London and Salisbury. The constitutional conference created a complicated electoral system establishing two levels of seats in the sixty-five-seat Assembly. Fifty seats were reserved for white legislators who were elected from an "A" roll of voters from which all but the best educated and most fortunate blacks would be barred. Fifteen seats were reserved for African legislators elected from a "B" roll of voters who met lower education, property, and income-level requirements. The effect of these two voter roll distinctions was that the one person, one vote hope of the NDP was not to be realized. The conference also established a Bill of Rights, but it was not retroactive so it did not change any discriminatory laws already in existence. Additionally, the conference did not address issues related to land, which contributed to the NDP's decision to reject the proposed constitution.

When the referendum was held in December 1961, the NDP and its membership boycotted the vote. Despite British Prime Minister Macmillan's commitment, as a result of the Monckton Commission report, to block any constitutional changes that did not have African support, the NDP's pleas to the British government to fulfill this commitment went unheeded. When the news of this perceived betrayal reached the broader NDP membership, violent protests ensued and the NDP party was banned on December 8, 1961.[29]

The Insurrection

On December 17, 1961, ten days after the banning of the NDP, Joshua Nkomo established ZAPU. Failing to win British support through legal and political means, the nationalist movement shifted its policy to engaging in more subversive activities in hopes that it would induce a British military intervention, to be followed by the imposition of a constitution deemed acceptable to the nationalists.[d] At the very least, the intent was to use violence to change attitudes of whites in

---

[d] The expectation of a British military intervention may have been based in part on the recommendations made by the Advisory Commission on the Review of the Constitution of Rhodesia and Nyasaland in 1960, which stated that the use of force would likely be required to hold the CAF together. Nationalist leaders at the time may have inferred the British government's willingness to intervene in political matters and to stave off potential instability in Northern Rhodesia (Zambia) and Nyasaland (Malawi) as an indicator that they would do the same in Southern Rhodesia, without consideration for the constitutional distinction of Southern Rhodesia as a self-governing colony and Northern Rhodesia and Nyasaland as protectorates.

Rhodesia and opinion within the British government. Addressing the decision to use political violence, one nationalist leader in 1960 noted:

> Although not for the purpose of guerrilla warfare but the purpose of carrying out acts of sabotage which were considered relevant to bring forth fear and despondency to the settlers of Rhodesia and to influence the British Government and the settlers in Rhodesia to accede to the popular revolutionary demands of the people in Zimbabwe.[30]

The current government in Rhodesia at the time was led by the United Federal Party, whose leader was Sir Edgar Whitehead. The United Federal Party was a pro-CAF party that sought full independence from Britain without immediate forfeiture of white political control to the nationalists. Concessions made to Britain in the drafting of the 1961 Rhodesian Constitution, which expanded black franchise, albeit only slightly and not to the acceptable levels nationalist leaders sought, gave Whitehead an ill-founded confidence that he could succeed in his mission of independence without majority rule.

Whitehead based his campaign on a platform that promised increased roles and rights for blacks in politics and society while seeking independence for Rhodesia without an automatic transition to majority rule. In an effort to win black support, Whitehead abandoned controversial laws being considered to further marginalize black political activity, modified a law that allowed black trade unions, prohibited race from being a factor in future wage negotiations, opened mid-level civil service positions to blacks, and opened public swimming pools to whites and blacks.[31] Nkomo is reported to have responded to the swimming pool desegregation with the remark, "We don't want to swim in your swimming pools. We want to swim with you in parliament."[32] Whitehead was still hoping for a partnership with the leadership of the black nationalist movement so that he could demonstrate to Britain that the Rhodesian black majority was content to wait fifteen or twenty years for racial parity in the halls of government. He had not anticipated the degree of the white community's backlash to his agenda.

The 1962 election was narrowly won by Winston Field's Rhodesian Front (RF) party. The RF party was dedicated to defending the status quo vis-à-vis white–black divisions of political, economic, and social power. In September of that year, the government responded to the growing security crisis by implementing emergency legislation that broadened the authority of the police and military and banned ZAPU, thereby driving its supporters underground.[33]

As the disintegration of the CAF became a reality in 1963, and it was clear Northern Rhodesia and Nyasaland would soon be granted independence, the RF leadership continued to insist that (Southern) Rhodesia should be granted independence coincidental to the date of the federation's official end. In terms of relations with the black population, the RF position was that the Land Apportionment Act of 1930 should not be changed, forced integration was unnecessary, and there was no cause to be hasty in preparing the black population for political responsibility. Playing on the fears of the white population, RF leaders depicted Whitehead as soft on law and order.

Because of his inability to convince Britain to take the nationalists' agenda more seriously, Nkomo decided to concentrate on gaining international recognition and support. However, his increased focus on the international community, coupled with recent criticism of his public support for the draft constitution, caused a segment of the movement to question his commitment to the cause of majority rule. This eventually led to an internal revolt that culminated with Nkomo suspending Robert Mugabe, Ndabaningi Sithole, and a handful of other leaders from the party in the summer of 1963.[34] As a result, on August 9, 1963, Sithole, Mugabe, and their supporters split from ZAPU and founded ZANU, with Sithole as its leader.

The premiership of Winston Field was relatively short but eventful in its quest to gain independence from Britain on terms satisfactory to Rhodesia's white minority. Field visited London to seek independence for Rhodesia, yet the British were prepared to grant independence only on the condition that Rhodesia made progress on implementing majority rule, that it ended racial discrimination, and that the terms of independence were agreeable to the entire population, both black and white.[35] These conditions proved unacceptable to the Rhodesian government, yet Field was not prepared to consider a Unilateral Declaration of Independence (UDI) as a viable option.[36] In April 1964 the party removed Field from his post as prime minister and installed his deputy, Ian Smith, a man with no reluctance to go as far as necessary to preserve minority rule.

On June 22, 1964, Nkomo, Sithole, and Mugabe, along with a handful of other activist leaders, were arrested for conspiring against the government, and they would remain incarcerated for a decade. With the political leaders of ZAPU and ZANU in jail, the conduct of the insurrections fell to lieutenants, many of whom had little leadership experience. In the mid-1960s, ZAPU and ZANU both established military wings: in the case of ZANU, the party established the Zimbabwe African National Liberation Army (ZANLA) in 1964, while ZAPU created an armed wing in 1965, which would eventually become the Zimbabwe People's Revolutionary Army (ZIPRA) in 1971. With the majority of the political leadership either in prison or abroad, the military components set out to establish and organize their guerrilla campaigns.

Before his incarceration, Nkomo had established a considerable degree of international recognition and support that would serve ZAPU well during its early years.[37] Organizations such as the Pan African Movement of East, Central, and Southern Africa (PAFMECSA) and the OAU enabled ZAPU to receive political and material support from the international community. The newly independent state of Zambia, under the leadership of Kenneth Kaunda, provided much-needed sanctuary to establish training and staging bases. The creation of the OAU and its Liberation Committee in 1963 provided a venue for PAFMECSA to channel resources to various liberation movements in Africa. Despite persistent pressure from regional states and the OAU, ZAPU and ZANU could not find common ground on which to reunify.

Rhodesian Unilateral Declaration of Independence

On November 11, 1965, Smith followed through on his threat to Britain and issued a UDI for Rhodesia.[e] For Smith, UDI represented an effort to "preserve justice, civilization and Christianity."[39] However, for many nationalists, UDI confirmed their view that a peaceful political solution to the problem of attaining majority rule was unattainable. With UDI came a new sense of commitment from the Smith government that neither Britain nor anyone else was going to force Smith to turn over control of Rhodesia to African majority rule before he decided it was time to do so.

Through a series of negotiations between the Rhodesian and British governments and between the Smith government and nationalist movement leaders over the course of thirteen years after UDI, proposal after proposal was rejected because of unacceptably long schedules for a move to majority rule or proposed voting arrangements that would have left significant power in the hands of the white minority. Each failed negotiation or attempted British intervention led to the further entrenchment of the nationalist position that victory would be attained only through a successful armed struggle.

---

[e] After UDI, the Soviet Union issued a statement noting that "the colonialists have committed a new crime against the African peoples. On November 11 the racialist regime of Ian Smith proclaimed the 'independence' of Southern Rhodesia. These actions are aimed at perpetuating in Southern Rhodesia a colonial system based on inhuman oppression of the Zimbabwe people, four million strong, by a handful of racialists . . . the South Rhodesian racialists would not have dared to carry out their criminal plans without a deal with the colonialists, who have permitted the racialist regime in Salisbury to acquire economic and military strength and who have rendered it all-out support. Nor could this crime have taken place without the blessing of other NATO countries, and in the first place the United States of America. The creation of yet another center of racialism—this time in Southern Rhodesia—is part of the overall plan of imperialist circles to erect an obstacle in the way of national liberation movements of the African peoples, the waves of which are drawing nearer and nearer to the last bulwark of colonialism. The Soviet government fully shares the view of the independent African states, expressed in decisions of the Organisation of African Unity, that the ruling circles of Britain will never be able to escape responsibility for this crime against the African peoples, for the national tragedy of the Zimbabwe people, who for many years now have been waging a stubborn struggle for their rights."[38]

The Impact of International Sanctions

The international community imposed sanctions on Rhodesia after the November 1965 UDI, and the sanctions remained in place until 1979. Rhodesia had actually been warned as early as October 1964 that a declaration of independence would be met with serious consequences. This advance notice provided an opportunity for the Rhodesian economy to plan for the eventuality of such an event.[40]

Sanctions were initially imposed on a gradual sliding scale. Initially, the British government restricted financial aid, export credit guarantees, and access to capital markets; banned sugar and tobacco imports; and implemented an embargo on oil imports.[41] This was followed shortly thereafter by United Nations (UN) Security Council Resolution 217, which called member states to "desist from providing it [the Rhodesian government] with arms, equipment and military material, and to do their utmost in order to break all economic relations with Southern Rhodesia, including an embargo on oil and petroleum products."[42] The tenor of Resolution 217 suggested that the sanctions were optional because the requesting state, Great Britain, refused to concede that the Rhodesian situation constituted "a threat to international peace and security," which was the grounds for mandatory economic sanctions under Chapter VII of the UN Charter.[43]

In December 1966, the sanctions were extended with the adoption of Security Council Resolution 232. While these sanctions remained selective, focusing on strategic materials, economically significant exports, and munitions, the UN Security Council made them mandatory for all UN member nations.[44,f] It was not until May 1968, nearly two and a half years after UDI, that the UN sanction regime was made comprehensive and mandatory with Security Council Resolution 253.[45] During this extended escalation of the sanction process, Rhodesia successfully adopted internal economic policies and leveraged the generosity of South Africa and the colonial regimes in Angola and Mozambique to minimize the worst effects of the sanctions.

The immediate impact of the sanctions was most acutely felt in the agricultural sector of the economy. The immediate ban on exports of

---

[f] The selective mandatory sanctions included asbestos, iron ore, chrome, pig iron, sugar, tobacco, copper, and animal products as well as military equipment, aircraft, motor vehicles, and petroleum. The comprehensive sanctions that would follow in 1968 were intended to complete the economic isolation of the Smith government.

sugar and tobacco to Great Britain impacted 71 percent of the value Rhodesia's exports to the former mother country. The export limitations to Britain were furthered expanded to include copper, chrome, asbestos, iron, steel, maize, and beef in December 1965.[46] This expansion encompassed 95 percent of the value of Rhodesia's exports to Britain. For white farmers, the solution was twofold. First, the government provided significant subsidies and a guaranteed market to tobacco farmers. Second, many farmers increased livestock production and expanded their agricultural repertoire to include irrigated wheat, maize, cotton, groundnuts, sugarcane, and sorghum.[47] These actions provided a basic income for farmers, allowing them to continue to operate, and they increased the self-sufficiency of the Rhodesian economy.

For black Africans in Rhodesia, the situation was not as bright. Those living on TTL (established as native reserves in the 1923 Constitution and renamed TTL in the Land Apportionment Act of 1930) had seen their lands become more crowded and overfarmed, limiting the available land for subsistence farming and grazing. For those who would have previously sought to emigrate from the TTL to urban areas or to white farm regions to provide labor, the economic restrictions due to the sanctions reduced their opportunities for agricultural employment, as well as reduced wages for those who could find employment. This phenomenon became worse as the years passed.

Despite the economic cost borne by the black African population, once the black nationalist leadership accepted that a negotiated settlement for majority rule was not a realistic option in 1972, they did not advocate for the dropping of sanctions. Testifying before a US congressional hearing in 1973, Eddison Zvobgo, director of the External Commission of the United African National Council (UANC), stated:

> It is not us who need sheets to sleep on or cars to come into the city, or spare parts to run the industries. We do not own the economy. Those comforts, which have been siphoned off by the sanctions, are totally irrelevant to the African people. So to suggest that sanctions hurt the Africans and therefore in the interest of the African we ought to drop the sanctions is nonsense.[48]

In contrast to the economic situation in the agricultural sector, in manufacturing and other nonagricultural occupations, African workers gained more than 150,000 jobs between 1965 and 1973.[49] The sanction-stimulated policy of import substitution and the side benefit of protection from foreign industries meant that manufacturing played an increasing role in the Rhodesian economy. Between 1965 and 1973, manufacturing output climbed from less than R$400 million to more than R$900 million, while agricultural output for the same period rose from less than R$200 million to nearly R$400 million.[50] Because of export limitations, the majority of manufacturing was oriented to local consumer goods, with particular geographic focus on Salisbury, the capital, and Bulawayo, the second-largest city in Rhodesia. As the war intensified after 1973, and the creative policies and practices of the government and industry leaders exhausted further avenues for economic expansion, the costs of the war and the effects of the sanctions became more evident to the government and the populace.

One of the most important and flaunted elements of the sanction regime was the prohibition of exporting oil and petroleum products to Rhodesia. Rhodesia relied heavily on its pipeline from Umtali, Rhodesia, to Beira, Mozambique, for its supply of oil. Despite strong rhetoric from the British government and support from the UN Security Council to prevent the landing of oil tankers in Beira as early as April 1966, the Rhodesian government found ways to continue the flow of refined oil products to Rhodesia virtually unabated for the first ten years of the sanction regime. The British blockade of the port of Beira, at a cost of more than £100 million over ten years, simply moved the supply chain south to Lourenço Marques (now Maputo), Mozambique.[51] Discussions between the British Labour and Conservative Parties regarding whether to extend the blockade to the entire Indian Ocean coast of southern Africa were a nonstarter. Such a move would have also severely impacted the flow of oil to South Africa, which no one in the British government, Labour or Conservative, was willing to entertain as a bearable cost to cutting off the Smith regime's access to oil.[52] The port at Lourenço Marques became the main supply point after the blockade of Beira. Oil arriving at the port was either shipped by railcar to Rhodesia or pumped via pipeline to South Africa where it was loaded onto railcars to be moved to Rhodesia. The list of oil companies participating in this sanction-busting activity included Britain's own British-Dutch Shell and state-owned British Petroleum, as well as Mobil, Caltex, and Total.[53]

The last significant issue related to the sanctions regime was the US Congress's adoption of the Byrd Amendment in November 1971. This action not only created international disagreement but also provided the Smith regime with much-needed foreign currency after the British closed their capital markets to Rhodesia immediately after UDI. One of the key provisions of the UN sanctions regime was the prohibition of importing strategic minerals from Rhodesia. In 1968 new deposits of nickel and chrome were discovered in Rhodesia.[54] Concerned that the Soviet Union was gaining a strategic advantage by purchasing strategic ore from Rhodesia in violation of international sanctions, the US Congress amended its Strategic Materials Act by passing the Byrd Amendment. This amendment forbade any prohibition on the US importation of any strategic and critical material from any non-Communist country so long as the importation of such material from Communist countries was not prohibited.[55]

Although Rhodesian chrome was not specifically mentioned in the legislation, US Senator Harry F. Byrd Jr., the namesake of the amendment, had made no secret of his long-standing opposition to the US and UN policies toward Rhodesia and his intention that the amendment was designed to facilitate transactions involving strategic materials from Rhodesia. Between 1972 and 1977, the United States imported $212 million worth of Rhodesian chrome, nickel, asbestos, copper, and other ores and alloys.[56] President Jimmy Carter successfully lobbied for the repeal of the Byrd Amendment, and the economic effect on Rhodesia was devastating. Not only was the existing ban on importation of Rhodesian ores fully implemented, but the ferrochrome and stainless steel alloys associated with Rhodesian exports to other countries were also subjected to sanctions. This act significantly cut the foreign capital supply to Smith's regime, which ultimately undermined Rhodesia's ability to finance the war.[57] Observers of economic sanctions also point out that the United States' move to abide by the sanction structure provided evidence of how effective sanctions can be if seriously implemented and enforced. The Byrd Amendment was repealed in March 15, 1977.[g]

It is difficult to make a definitive assessment of whether sanctions were a pivotal causal factor in the demise of the Smith regime. The Smith government's creative policies to incentivize farmers and industry to diversify their activities to more aptly suit the conditions of an

---

[g] Public Law 95-12 gave President Carter the authority to halt U.S. imports of chrome from Rhodesia. This legislation passed the House of Representatives by a 250–146 vote on March 14, 1977, and passed the Senate by a 66–26 vote on March 15, 1977.

isolated economy, combined with the willingness of major economic players in the region and the world to ignore key provisions of the sanction regime, particularly in the area of oil and strategic ores, allowed the white population to weather initial hardships. Because most of the black population did not consider the sanctions overly relevant to their place in Rhodesian society or the economy, the sanctions were largely irrelevant to the decisions nationalist leaders made about the political or military conduct of the struggle. It is clear that as the guerrilla campaign continued into the mid- to late 1970s, and as South Africa vacillated in rendering aid to Rhodesia's counterinsurgency operations,[h] the sanctions exacerbated the economic challenges Rhodesia faced to continue its fight against the increasingly inevitable outcome of black majority rule.

Armed Conflict and Political Maneuvering

Over the course of the decade after UDI, Rhodesia would become engulfed by two separate and competing insurrections. During that time, both ZAPU and ZANU underwent significant evolution and transformation. The period of 1965–1970 could be described as a series of unsuccessful military endeavors that yielded little to no benefit for the movements. The string of failed guerrilla operations resulted in the eventual reassessment of strategies and tactics on the part of ZAPU and ZANU and their respective Soviet and Chinese sponsors. The new campaigns that materialized as a result of these strategic reassessments were a reflection of the influence and perspectives of Soviet and Chinese understanding of revolutionary warfare.

As the conflict intensified into the 1970s, the insurgents and Rhodesian Security Forces (RSF) both demonstrated considerable skill in adapting their tactics. In particular, the Rhodesian forces achieved a well-deserved reputation for innovative and daring counterinsurgency

---

[h] The Vorster government in South Africa curtailed its military and police support to Rhodesia as part of its détente initiative in 1975-77, partially restoring its aid after the failure of the Geneva peace talks. After the Botha government came to power in 1978, South Africa increased its military support to Rhodesia through the end of the conflict.[58]

tactics.[i] Additionally, the abilities of the insurgents, both ZAPU and ZANU, to sustain a long succession of tactical losses demonstrated their organizational resilience and commitment. Despite all of ZAPU's efforts to garner international recognition to the cause, by 1977 ZANU had established itself as the dominant insurgency posing a threat to the Rhodesian government.

Behind the scenes of the ZAPU–ZANU competition for control over the voice of the liberation struggle was a dance between South Africa and the "Front Line States"[j] of Zambia, Botswana, Mozambique, Angola, and Tanzania. As the conflict wore on patience grew thin because of the economic costs all parties were paying for the military stalemate in Rhodesia. Through a series of negotiations conducted over a roughly two-year span from the summer of 1972 to the fall of 1974, Zambian President Kenneth Kaunda and South African President B. J. Vorster worked to bring all parties to the table to find a peaceful path forward. Although Vorster was initially hesitant to agree to some of the terms, such as the immediate withdrawal of the South African Police from Rhodesia, the results of the military coup d'état in Portugal in 1974 (leading to the independence of Mozambique and Angola) convinced him that a military victory for Smith was unlikely. Instead, he believed South Africa would be better off having some influence over the formation of a new Rhodesian (or Zimbabwean) regime than risking a Soviet-backed government that would provide support to the South African liberation movement, the African National Congress (ANC).

South Africa persuaded Smith to release key liberation leaders in detention, including ZAPU's Nkomo and ZANU's Sithole and Mugabe, so that they could attend a constitutional conference in Lusaka intended to set a timetable for establishing majority rule. Throughout

---

[i]   An example of these tactics includes operations in which specially trained Rhodesian forces, the Selous Scouts, posed as guerrillas to infiltrate guerrilla territory and positions. Additionally, the security forces made great use of small, highly mobile forces to find, pursue, and block guerrilla forces through the use of ground tracking and parachute operations.

[j]   After the political earthquake caused by the 1974 coup d'état in Portugal, which led to independence for Angola and Mozambique, South Africa initiated a policy of détente in an effort to reach a negotiated settlement to the conflict in Rhodesia. In response to this effort, the regional countries of Zambia, Botswana, Mozambique, and Tanzania attempted to develop a common policy for negotiations with Rhodesia, and this grouping came to be known as the Front Line States.[59] See chapter 5 for more information on the role of the front line states in the Rhodesian conflict.

the fall of 1974, preparations were made for the conference in Lusaka, and the liberation groups were represented by Nkomo (ZAPU), Sithole (ZANU), Bishop Muzorewa (UANC),[k] and James Chikerema (FROLIZI, the Front Line for the Liberation of Zimbabwe).[l] On December 8, 1974, the four leaders signed a unification agreement that called on Smith, among other things, to release all political detainees and lift the ban on ZAPU and ZANU. Muzorewa became the compromise chairman, and it was agreed that a UANC congress would be held in the spring of 1975 with the goals of adopting a new UANC Constitution and electing new leadership. In reality, the cease-fire was very short-lived, and Smith stopped the release of prisoners in January 1975.

This latest failure at a peaceful solution brought about the direct intervention of the Front Line States. Continued discussions between the four liberation groups were unsuccessful in establishing an agreed-on structure or power-sharing arrangement under the UANC umbrella. The liberation group leaders and Front Line States put negotiations on the back burner and escalated the war effort. As a condition of using bases in his country, Mozambican President Samora Machel required that the war be fought as one army, not two.[60] The new unified army was named the Zimbabwe People's Army (ZIPA), to be led by a combination of ZANLA and ZIPRA leaders. While full integration of the respective insurgent armies was never realized operationally, their operating strength both inside and outside Rhodesia grew rapidly from 1975 onward. By 1977, an estimated 4,000 guerillas were operating within Rhodesia's borders, the preponderance of these belonging to ZANLA.[61, m]

Politically, however, the calculation was somewhat different. By late 1976 it became obvious that the ZIPA experiment had not worked as hoped. One reason for the failure of the military unification was the lack of political union. In anticipation of renewed round of negotiations sponsored by the United States and Britain, the Front Line States

---

[k]   The UANC formed in 1971 to oppose a potential political settlement being considered by Rhodesia and Britain. For more information, see the "Political Activities" section of chapter 4.

[l]   FROLIZI was a ZAPU splinter group Chikerema created in October 1971. It did not become a key player, despite Chikerema's efforts to present it as a unified movement of dissatisfied ex-ZANU and ex-ZAPU figures.

[m]   The number of ZIPRA personnel operating within Rhodesia by the end of 1977 was estimated at about five hundred.[62]

encouraged ZANU and ZAPU to bury their differences and agree upon a united negotiating position under the auspices of the Patriotic Front (PF).[63] At the same time, the US and South African governments pressured the Rhodesian government to pursue a negotiated settlement. This initiative was heralded by a speech delivered by US Secretary of State Henry Kissinger which asserted the US government's support for transition to black majority rule in Rhodesia, restated its intention to revoke the Byrd Amendment, and pledged "unrelenting opposition" to white minority rule "until a negotiated settlement is achieved."[64] Under pressure from the South Africans to yield to Kissinger's demands, suffering economically and financially from the loss of foreign exchange to pay for the war, forced to dig deeper into available white manpower resources (thereby negatively impacting the economy)[65] and beginning to suffer from white emigration, on September 24, 1976, Smith conceded the principle of majority rule, "provided that it is responsible rule." From that point on, Smith sought constitutional guarantees to preserve white influence in the legislature and continued dominance over key elements of the executive, primarily the military and internal security functions.

The Kissinger initiative effort culminated in the Geneva talks held from October 28 through December 14th 1976, which included the British government, the African nationalists, and the Smith delegation.[66] In these talks, Nkomo and Mugabe shared the PF seat. No resolution to the conflict was achieved at Geneva. The Smith delegation and the liberation groups disagreed on the timetable for majority rule; the groups offered to give Smith twelve months, but Smith insisted on the two year interim period he had announced in September.[67] Additionally, Smith demanded that his RF party retain control of the key Ministries of Defense and of Law and Order during the transitional period.[68] The Africans insisted that they control the interim government. Smith rejected this demand.

The failure of the Geneva talks did not bring an end to efforts at a negotiated settlement. Following the onset of the Carter Administration in 1977, subsequent British-American proposals focused on a transfer of power to a transitional government under a British caretaker commissioner in preparation for majority rule under a new constitution. The proposals also advocated that, immediately after the negotiation of a cease-fire, the Rhodesian army be integrated with the guerrilla forces. These proposals were unacceptable to Smith, who continued to insist that constitutional limitations on the franchise be preserved, that white investments and real property rights be protected in a new constitution, and that the guerrillas be excluded from post-independence military and police establishments.[69] With the war continuing to escalate throughout 1977, Smith turned to moderate African leaders who had disassociated themselves from the guerrillas in pursuit of an internal settlement, culminating in a March 1978 agreement with Muzorewa, Sithole, and Chief Jeremiah Chirau, a prominent Shona chief. This agreement called for an Executive Council consisting of Smith and the three black African leaders and a Ministerial Council of eighteen that would consist of one black and one white co-minister for each of the nine governmental departments.[70] Additionally, twenty-eight seats within the one-hundred-seat Parliament would be reserved for whites, denying the black population a three-quarters majority required for some constitutional changes, and the twenty-eight seats would be guaranteed for whites for at least ten years. Additionally, the civil service and defense and police forces would be "maintained in a high state of efficiency and free from political interference," which meant that the majority of whites in the bureaucracy would keep their positions.[71]

After a one-year transition period, the first majority-rule elections were held in May 1979, with Muzorewa's UANC party winning a clear majority of fifty-one seats.[72] The new government, under the new national name Zimbabwe–Rhodesia, failed to garner international recognition nor an end to the economic sanctions that were now crippling the country. While the Thatcher government in Britain was more favorably disposed toward Muzorewa's government than its predecessor, the Carter Administration retained economic sanctions despite a US Senate resolution declaring that the May 1979 elections were free and fair, and that conditions for black majority rule had been met. The African nations threatened a trade boycott if the UK recognized the Muzorewa government. Under these pressures, the Thatcher government continued to seek a comprehensive settlement on the same lines as before.[73]

Additionally, while efforts at reaching a political settlement proceeded, military operations of both the insurgents and the Rhodesian government continued to rage onward. The PF regarded Muzorewa as a stooge, and Ian Smith sought to grind down ZAPU and ZANU through attacks on insurgent camps in Mozambique, Zambia, and Angola in support of a negotiated settlement with moderate black leaders, while at the same time impressing these moderates with white military power.[74] Toward this end, Rhodesian forces launched a number of large-scale military operations in the late 1970s, such as Operation Dingo, which featured an air assault on ZANLA headquarters near Chimoio in Mozambique that involved almost the entire Rhodesian Air Force, and a follow-up assault on a ZANU camp in Tembue.[75] The camp at Chimoio held an estimated nine thousand ZANLA personnel, while that at Tembue held four thousand, and the number killed and wounded was estimated in the thousands (ZANU claimed that many of the casualties were women and children, as the Chimoio camp also housed schools and hospitals).[76]

In October 1978 the Rhodesian Air Force bombed guerrilla camps in Chikumbi and Mkushi, both in Zambia, and in April 1979 Rhodesian forces attacked the ZIPRA military command, located in Lusaka, with the intention of killing Nkomo.[77] Additionally, within forty-eight hours of his accession to power, Muzorewa authorized attacks on neighboring countries, and Rhodesian forces attacked ZAPU's intelligence headquarters located in a suburb of Lusaka.[78] Furthermore, shortly before the Lancaster House talks began in the fall, Rhodesian forces attacked ZANLA bases near Aldeia de Barragem in Mozambique, and they proceeded to attack the economic infrastructure of neighboring countries. For instance, in October and November 1979 Rhodesian forces attacked Zambia's rail and road network, and in September they attempted to destroy much of the transportation infrastructure in Gaza Province in Mozambique. The intention was to stop the infiltration of guerrillas and supplies into Rhodesia and to compel regional countries to pressure ZANU and ZAPU to take a more moderate approach in the Lancaster talks.[79]

Yet, by this time, the insurgents were well entrenched inside the country. By the start of the Lancaster talks ZANLA forces within the country exceeded twenty thousand, and the government had started to lose control of rural areas.[80] Additionally, the insurgents succeeded in carrying out a number of terrorist and sabotage operations, including the bombing of a Woolworth's store in Salisbury and the destruction of the fuel depot in the city.[81] Insurgents also repulsed a September 1979 Rhodesian assault on a ZANLA control center and the headquarters of the Frente de Libertação de Moçambique (Mozambique Liberation Front, or FRELIMO) 2nd Brigade in Mapai, Mozambique, and by this time ZIPRA preparations for a large conventional assault throughout Rhodesia, known as Operation Zero Hour, were well under way.[82]

Resolution: Liberation and Transition of Government Power

Negotiations proceeded while fighting raged. At a British Commonwealth meeting in Lusaka in August 1979, the new British government led by Margaret Thatcher was pressured to convene an all-parties constitutional conference to seek an end to the civil war. The result was the Lancaster House Conference held in London in the fall of 1979. This conference resulted in the Lancaster House Agreement of December 1979, which ended the Rhodesian Bush War. As part of the agreement, Smith was to hand over control of the government to a British high commissioner who would oversee preparations for elections to be held in the spring of 1980. The agreement also called for demobilization of the guerrilla forces and the wartime footing of the RSF. According to the plan, the cease-fire was to take effect on December 28, 1979, and all guerrillas were to be gathered at rendezvous points so that they could be transported to sixteen assembly areas in Rhodesia for disarmament and demobilization.[83] However unrealistic that timetable was given the breadth of territory in which guerrillas were deployed, it was obvious after the first week of the cease-fire that it was taking effect.

On March 4, 1980, elections were held to select the first genuine majority government in Rhodesian (Zimbabwean) history. In the new one-hundred-seat Parliament, with only twenty seats reserved for whites, Robert Mugabe's ZANU-PF party won a clear majority of fifty-seven seats. Nkomo's PF party won twenty seats, and Bishop Muzorewa won only three.[84] The election was plagued by accusations of intimidation perpetrated by Mugabe loyalists, but the pressure was too great for British elections commissioners to declare the results anything but free, fair, and reflecting the opinion of the electorate. Mugabe's initial overtures to the white population, and Ian Smith in particular, were gracious. He retained General Peter Walls, the white commander of the RSF, to preside over the integration process of building a new united Zimbabwean army. He offered white RF parliamentarians positions as leaders of economic ministries. Mugabe also made offers of key posts to Nkomo and his leading party members, most of which were rejected. The bitterness of the nearly two-decade-long competition between ZAPU and ZANU could not be swept under the carpet for the appearance of unity. By 1981 Mugabe was making public speeches outlining his plan for the creation of a one-party state. Following a period of conflict between ZANU and ZAPU in the wake of the 1980 election, ZANU gained a dominant position, forcing Nkomo to merge the two parties and join Mugabe's government in 1987, creating the ZANU-PF of today's Zimbabwe. Following Nkomo's death in 1999, ZAPU was reconstituted as an opposition party. Robert Mugabe ruled Zimbabwe for thirty-seven years, resigning as President of Zimbabwe on November 20. 2017.[85]

NOTES

1   Robert Blake, *A History of Rhodesia* (New York: Alfred A. Knopf, Inc., 1977), 4.

2   Ibid., 5.

3   Ibid.

4   J. R. T. Wood, "Rhodesian Insurgency," Rhodesia & South Africa website, accessed November 10, 2014, http://www.rhodesia.nl/wood1.htm.

5   Glenn V. Stephenson. "The Impact of International Economic Sanctions on the Internal Viability of Rhodesia," *Geographical Review* 65, no. 3 (1975): 380.

6   Ibid., 384.

7   Paul L. Moorcraft and Peter McLaughlin, *The Rhodesian War: A Military History* (Mechanicsburg, PA: Stackpole Books, 2008), 18.

8   M. Elaine Burgess, "Ethnic Scale and Intensity: The Zimbabwean Experience," *Social Forces* 59, no. 3 (1981): 605.

9   Ibid.

10 David Martin and Phyllis Johnson, *The Struggle for Zimbabwe* (London: Faber and Faber, 1981), 52.

11 Ibid., 53.

12 Ibid.

13 Blake, *A History of Rhodesia*, 198.

14 Ibid., 204.

15 Peter Baxter, *Rhodesia, Last Outpost of the British Empire 1890–1980* (Alberton: Galago Publishing, 2010), 49–56.

16 Blake, *A History of Rhodesia*, 42–61.

17 Baxter, *Rhodesia*, 50–51.

18 Blake, *A History of Rhodesia*, 47.

19 Ibid., 53.

20 Ibid., 198.

21 R. Cranford Pratt, "Partnership and Consent: The Monckton Report Reexamined," *International Journal* 16, no. 1 (1960/1961): 38.

22 Quoted in Larry W. Bowman, *Politics in Rhodesia: White Power in an African State* (Cambridge, MA: Harvard University Press, 1973), 28.

23 Pratt, "Partnership and Consent," 41–42; and Bowman, *Politics in Rhodesia*, 28.

24 Pratt, "Partnership and Consent," 41.

25 Eliakim M. Sibanda, *The Zimbabwe African People's Union 1961–1987: A Political History of Insurgency in Southern Rhodesia* (Asmara, Eritrea: Africa World Press, Inc., 2005), 34–35.

26 Bowman, *Politics in Rhodesia*, 45.

27 Ibid., 47.

28 Anthony R. Wilkinson, *Insurgency in Rhodesia, 1957–1973: An Account and Assessment*, Adelphi Paper No. 100 (London: The International Institute for Strategic Studies, 1973), 4.

29 Sibanda, *Zimbabwe African People's Union*, 61.

30 Michael Raeburn, *We Are Everywhere: Narratives from Rhodesian Guerrillas* (New York: Random House, 1978), 195.

31 Baxter, *Rhodesia*, 324.

32 Sibanda, *Zimbabwe African People's Union*, 53.

33 Nathan Shamuyarira, *Crisis in Rhodesia* (New York: Transatlantic Arts, 1965), 202–203.

34 Sibanda, *Zimbabwe African People's Union*, 91.

35 Peter Calvocoressi, *World Politics since 1945*, 9th ed. (Abingdon, Oxon, England: Routledge, 2009), 605.

36 Martin and Johnson, *Struggle*, 71.

37 William Cyrus Reed, "International Politics and National Liberation: ZANU and the Politics of Contested Sovereignty," *African Studies Review* 36, no. 2 (1993): 37.

38 Paul Halsall, "Rhodesia: Unilateral Declaration of Independence Documents, 1965," *Modern History Sourcebook*, July 1998, http://www.fordham.edu/halsall/mod/1965Rhodesia-UDI.html.

39 Ian Smith, *Bitter Harvest: Zimbabwe and the Aftermath of Independence: The Memoirs of Africa's Most Controversial Leader* (London: John Blake, 2008), 106.

40 William Minter and Elizabeth Schmidt, "When Sanctions Worked: The Case of Rhodesia Reexamined," *African Affairs* 87, no. 347 (1988): 211.

41 Stephenson, "Impact of Sanctions," 377.

42 United Nations Security Council, Security Council Resolutions, S/RES/217 (1965), https://undocs.org/S/RES/217(1965), accessed October 12, 2017

43 Minter and Schmidt, "When Sanctions Worked," 212.

44 Ibid. See also United Nations Security Council, Security Council Resolutions, S/RES/217 (1965), https://undocs.org/S/RES/232(1966), accessed October 12, 2017.

45 Stephenson, "Impact of Sanctions," 377. See also United Nations Security Council, Security Council Resolutions, S/RES/235 (1968), https://undocs.org/S/RES/253(1968).

46 Minter and Schmidt, "When Sanctions Worked," 212.

47 Stephenson, "Impact of Sanctions," 382.

48 Minter and Schmidt, "When Sanctions Worked," 232.

49 Stephenson, "Impact of Sanctions," 382.

50 Ibid., 383.

51 Minter and Schmidt, "When Sanctions Worked," 216.

[52] Blake, *A History of Rhodesia*, 392.

[53] Minter and Schmidt, "When Sanctions Worked," 217.

[54] Blake, *A History of Rhodesia*, 394.

[55] Minter and Schmidt, "When Sanctions Worked," 218.

[56] Ibid.

[57] Ibid., 219.

[58] Paul Moorecraft, "Rhodesia's War of Independence," *History Today* 40, no. 9, (1990), http://www.historytoday.com/paul-moorcraft/rhodesias-war-independence, accessed October 12, 2017.

[59] Reed, "International Politics and National Liberation," 42–43.

[60] Martin and Johnson, *Struggle*, 217.

[61] John Day, "The Rhodesian Internal Settlement," *World Today* 34, no. 7 (1978): 270.

[62] Ibid.

[63] Sibanda, *Zimbabwe African People's Union*, 181. See also Martin Meredith, *The Past is Another Country: Rhodesia 1890-1979* (London: Andre Deutsch, Ltd., 1979), 268.

[64] Michael T. Kaufman, "Chrome Ban Asked," *New York Times*, April 26, 1976. See also Department of State Bulletin, Volume LXXIV, No. 127, May 31, 1976, 672–679.

[65] Moorcraft and McLaughlin, *The Rhodesian War*, 148–149. Despite opposition from the commercial sector, in 1977 the conscription age was increased to include the 38–50 age cohort, while the maximum call-up for those under 38 was extended to 190 days per year.

[66] Meredith, *The Past is Another Country*, 242–293.

[67] Sibanda, *Zimbabwe African People's Union*, 213.

[68] Day, "The Rhodesian Internal Settlement," 269.

[69] Meredith, *The Past is Another Country*, 294-301; See also Day, "The Rhodesian Internal Settlement," 272–273.

[70] Day, "The Rhodesian Internal Settlement," 273.

[71] Ibid.

[72] Sibanda, *Zimbabwe African People's Union*, 215.

[73] Meredith, *The Past is Another Country*, 365–377.

[74] Moorcraft and McLaughlin, *The Rhodesian War*, 147, 162.

[75] Ibid., 150.

[76] Ibid.

[77] Ibid., 155, 159.

[78] Ibid., 163.

[79] Ibid., 163–164. South African forces were also heavily involved by this point in the conflict. Moorcraft and McLaughlin noted that by November 1979 South African forces were heavily present in the southeast, an in particular in the Sengwe TTL and along the border. The South Africans used artillery bombardment to create insurgent movement, and by December the South Africans were also operating in the area north of Chiredzi and were planning to put a battalion into each major operational area.

[80] Ibid., 168.

[81] Ibid., 81, 160.

[82] Ibid., 81.

[83] Ibid., 77, 164–165.

84 Martin and Johnson, *Struggle*, 321.

85 Norimitsu Onishi and Jeffrey Moyo, "Robert Mugabe Resigns as Zimbabwe's President, Ending 37-Year Rule," *New York Times*, November 21, 2017, https://www.nytimes.com/2017/11/21/world/africa/zimbabwe-mugabe-mnangagwa.html, accessed November 30, 2017; see also Martin and Johnson, *Struggle*, 330.

CHAPTER 4.
THE INSURGENCIES

THE RESISTANCE MOVEMENT

In the two decades after the Second World War, the British government gradually relinquished the remainder of its empire in Africa. From 1957 through 1964, except for Southern Rhodesia all of the former British colonies gained independence with black African majority governments. In response to African nationalist movements, the British first attempted to establish colonial federations in their colonies in east and central Africa. These federations were invariably opposed by the nascent African nationalist movements.[1] From 1953 to 1963, Southern Rhodesia belonged to the Central African Federation (CAF), corresponding to the current states of Zimbabwe (Rhodesia/Southern Rhodesia), Zambia (North Rhodesia), and Malawi (Nyasaland). It was during this period that the African nationalist movement took root in Southern Rhodesia, the CAF colony with the largest European minority.[2,a]

Before the dissolution of the CAF, the British sought to establish what they referred to as a racial partnership. This policy failed to elicit support from the black African majority, whose aspirations exceeded the political and economic benefits offered by the constitutional arrangements proposed for the federation in 1953 and 1961. While some African leaders supported separate voting rolls for blacks and whites, a single voting roll with a limited franchise, or a bloc of legislative seats in the federation Assembly, the African nationalist position gravitated to full independence with an unlimited franchise on the basis of one person, one vote.[4]

At the same time, the Southern Rhodesia economy was growing rapidly. In contrast to the rest of the CAF, small towns and cities throughout Southern Rhodesia were characterized by industrial growth that accompanied the colony's robust agricultural sector. Gross domestic product in real terms averaged over 10 percent annually from 1945 to 1953, and between 1953 and 1957 the number of manufacturing enterprises grew from 700 to 1,300.[5]

The agricultural sector was also growing, based largely on cash crops such as tobacco. The growth of the agricultural sector led to increasing tensions over the usage of land, which the Land Apportionment

---

[a] Peters states that the white minority made up 8 percent of the population of South Rhodesia, 3 percent of North Rhodesia, and 0.3 percent of Nyasaland.[3]

73

Act apportioned into four sectors: European, native, forest, and unreserved land. In the 1950s, 110,000 black Africans were expelled from farming areas reserved for Europeans. British attempts at agricultural reform were met by opposition from the black African majority, who were forced to sell their cattle stock for a pittance. Indeed, the Native Land Husbandry Act, intended to "control . . . the utilization and allocation of land occupied by natives to ensure its efficient use for agricultural producers and to require natives to perform labor for conserving natural resources,"[6] was deeply threatening to traditional African agriculture, which required sufficient land for herding. Colonial policy consigned black African farmers to marginal lands that were least connected to transportation networks.[7]

In both the white-owned rural areas as well as the townships surrounding cities such as Salisbury (Harare) and Bulawayo, the number of black African wage earners grew to half a million by 1954. With this growth, questions of economic privilege and the distribution of wealth arose among black African wage earners. Labor unions within Southern Rhodesia were segregated, pitting white laborers against black workers. Caucasians dominated skilled labor positions, and discriminatory practices prevented unskilled black and white workers from competing on an equal basis.[8]

The economic inequality between black Africans and white Europeans was exacerbated by inadequate educational opportunity for most black Africans and other non-white minorities in Southern Rhodesia. Prior to World War II, the best and most-funded educational venue available to black Africans was the network of mission schools throughout the Rhodesian countryside. The majority of black Africans received a rudimentary primary education in kraal schools, which received only 27 percent of government funding for African education. Government-run secondary schooling was unavailable to black Africans until the postwar period; in 1960, average spending per secondary school student was £8 for blacks as opposed to £103 for whites. Although the number of primary schools increased by 130 percent from 1953 to 1960, a 1962 census indicated that 40 percent of African children were not attending school and 47 percent of males and 59 percent of females born after 1947 had never been to school. With an annual population growth rate of 2.5 percent, this situation continued to worsen as the Rhodesian Front (RF) took power and slashed education funding for black Africans. The RF government also increased school fees, which increased the number of dropouts as more parents became unable to afford the new rates.[9]

Up through World War II, the growth of the white population in Southern Rhodesia was largely driven by immigration; for instance, during the period from 1901 to 1911, immigration accounted for 88 percent of the white population's growth, and from 1931 to 1941, it accounted for 58 percent of the growth, yielding a total white population of 68,954 in 1941.[10] For working-class British servicemen, the Rhodesian economy offered substantial benefits in the postwar environment, offering the chance for a new start in a society that—at least for the white minority—was more egalitarian than Great Britain.[11] Consequently, by 1961, the white population of Southern Rhodesia reached 221,000, increasing by 16 percent a year.[12] The Nationalist Party victory in South Africa in 1948 also triggered a wave of English-speaking immigrants from South Africa. With an Afrikaner minority of 15 percent of the white population, white Rhodesian society continued to hang on psychologically to the past, "conscious of its Britishness and [more] determined to preserve its corporate identity than those who had remained in the homeland."[13]

Such attitudes impeded the ability of the British government at both imperial and local levels to deal with rising racial and class tensions within Southern Rhodesia. As the black African majority grew more restive, the white European minority grew more defensive, less and less willing to seek a compromise. Increasingly repressive measures, including the banning of black African political organizations and the detention of black Africans, accelerated. The Southern Rhodesia African National Congress (SRANC) was banned in 1959. Its successor, the National Democratic Party (NDP), was banned in 1961, followed by the creation and subsequent banning of the Zimbabwe African People's Union (ZAPU) in 1962.[14] At the same time, black-on-black violence accelerated, as the African nationalist leaders sought to close ranks. Older and more moderate black African leaders were supplanted by angry young men. In 1958, the American Vice Consul Robert Murphy reported the following conversation with two of the leaders of the City Youth League (a precursor to the NDP), James Chikerema and George Nyandoro:

> Chikerema then stated flatly, "Bloodshed is unavoidable in this country." Nyandoro attempted to soften

> Chikerema's statement claiming, "But we can control our followers. We know that when we threaten the very livelihood of the European through attempting to change the present economic organization, they will try to shoot us." "And," said Chikerema, "then our people will rise."[15]

In this environment, the radical ideologies of the left found fruitful soil. The 1950s and early 1960s saw the height of the Cold War, a time of worldwide ideological struggle between liberal democracies and Communist states. Southern Rhodesia remained something of a colonial backwater, lagging economically behind its larger and more modern neighbor to the south and lagging politically behind its less developed neighbors to the north. As both the black African majority and the white European minority radicalized during the crucial years of British decolonization, the fate of Southern Rhodesia remained unresolved, while every other British colony gained independence. In the meantime, the Cold War protagonists—the United States, the Soviet Union, and, after the Sino-Soviet split, Communist China—warily moved in, seeking advantage in the global struggle.[16]

For a brief period during the Eisenhower and Kennedy administrations, the US government maintained a cordial and supportive relationship with the African nationalist leadership in Southern Rhodesia. The American vice-consul in Salisbury, Edward Mulcahy, was actively engaged with these leaders, inviting ZAPU functionaries Joshua Nkomo and Robert Mugabe to his residence. As Mulcahy recounts, "They were often at my house, drinking my beer. Robert preferred my Scotch."[17] The American Federation of Labor and Congress of Industrial Organizations (AFL-CIO) used its influence in the Brussels-based International Confederation of Free Trade Unions to channel money to the Southern Rhodesian Trade Union Congress, a black African trade union Nkomo and Reuben Jamela established in 1953. As both the black African trade unions and the African nationalist political organizations split in early 1960, Nkomo was criticized within African nationalist circles for taking American money. He responded by attacking both the Americans and (following the establishment of ZANU) his ZANU adversaries, accusing them of selling out to American capitalist interests. In fact, ZANU officials did meet with US State Department officials in Washington and Dar es Salaam, seeking financial support. Although the US government does not appear to have given any direct financial support to ZANU, it did consider the possibility of using its influence to encourage private contributors to support ZANU—until ZANU officials attacked the US publicly in the Chinese media.[18]

Before the ZANU/ZAPU split, the African nationalist leaders had resolved to initiate the armed struggle against the Rhodesian government. By the time of the Rhodesian Unilateral Declaration of Independence (UDI) in 1965, ZANU and ZAPU cadres had been training in the Soviet Union, China, Egypt, Ghana, and Tanzania, in some cases for at least two years.[19] According to records in the State Archive of the Russian Federation, Southern Rhodesian African nationalist contacts with the Eastern Bloc began in April 1960, at an Afro-Asian People's Solidarity Organization (AAPSO) meeting held in Beirut, and were followed by a visit to Czechoslovakia where NDP representatives submitted a request for "special training (security, defense)."[20] The NDP also requested and received financial aid, arguing that it was "conducting certain work in . . . Katanga . . . in defence of the lawful Congolese government of P. Lumumba."[21,b] Discussions continued between ZAPU and the Soviet government throughout 1962 and 1963, culminating in a ZAPU request for military training. The first set of ZAPU cadres sent to China in 1962 split between the two organizations upon returning to Tanzania.[23]

As both ZAPU and ZANU radicalized, turning away from the West, their leaders embraced Soviet and Chinese political and military doctrine more strongly. The Sino-Soviet split, in addition to personal ambition and tribal differences, drove these two groups further apart and made them more dependent on their respective Communist sponsors. As Table 4-1 illustrates, the ZANU/ZAPU split was not unique to Rhodesia–Zimbabwe; all across southern Africa, resistance movements arose, often divided along tribal as well as ideological lines, with competing state sponsors lining up to support them.[24] As revealed in KGB archival material smuggled into Great Britain by a Soviet defector, Soviet intelligence agents used the opportunity to train ZAPU cadres as a means of recruiting them to the KGB. Of the nine penetrations reported, two became full-fledged agents of the KGB. Nkomo himself was described by the Soviets as a "bourgeois nationalist."[25]

---

[b]   The State Archive of the Russian Federation records show that the NDP received $8,400 from the Soviets.[22]

Table 4-1. Alignment of state-sponsored resistance groups in southern Africa.

Target Nation	Russian-Sponsored Groups	Chinese-Sponsored Groups
Angola	Movimento Popular de Libertação de Angola (People's Movement for the Liberation of Angola, MPLA)	National Union for the Total Independence of Angola (UNITA)
Rhodesia (Zimbabwe)	ZAPU	ZANU
Mozambique	Frente de Libertação de Moçambique (Mozambique Liberation Front, FRELIMO)	Revolutionary Committee of Mozambique (COREMO)
Southwest Africa (Namibia)	South West Africa People's Organization (SWAPO)	South West Africa National Union (SWANU)
South Africa	(South African) ANC	Pan-African Congress (PAC)

ᵃ The Chinese supported COREMO until that organization's demise in the early 1970s, thereafter supporting FRELIMO, which had a pro-Soviet and a pro-Chinese wing.[26]

Much of the external governmental support for the two Zimbabwean resistance movements was channeled through the auspices of the Organisation of African Unity (OAU). In addition to sanctuary provided by Zambia, Tanzania, and (after 1974) Mozambique, OAU states provided weapons, training, and access to radio broadcasting facilities, as well as financial contributions through the OAU Special Fund for Liberation. Both the OAU and the Soviets encouraged collaboration, with mixed results.[27]

Nature of Resistance Movement or Insurgency

Throughout the balance of the Rhodesian conflict, even to today, a debate has raged as to the scope and intention of the African nationalist insurgency movements. As ZAPU and ZANU turned to the Communist world for support, their rhetoric more and more incorporated the Communist critique of capitalism and imperialism. The subsequent actions of the Zimbabwean government after independence in 1979 to some extent sustained the white Rhodesian minority's allegations that these movements were inspired by Communist—or at least Socialist—ideology with a twist of racial revanchism, but the white minority's fear of political and economic expropriation was only realized gradually in the subsequent years. Were ZAPU and ZANU in fact African nationalist movements seeking reformist goals through violent means, or were they full-blooded Communist conspiracies as the RF government alleged?[28] One Marxist source, comparing the post-war ZANU government's performance in relation to its wartime rhetoric, concluded:

> The leadership of the nationalist movement in Zimbabwe was never able to fight consistently against imperialism. On the contrary, the leadership always attempted to contain the struggle of the African people. As the anti-imperialist movement developed, the leadership understood that its interests could be defended only through a compromise with imperialism. This was not merely the result of the narrow class interests of the petit bourgeois leadership. The political programme of the nationalist movement was a clear expression of the petit bourgeois politics dominating the movement as a whole. The inherent political weakness of the nationalist movement's programme meant that there was always the danger that the tendency to seek a compromise with imperialism would actually occur in practice.

> Today, if the struggle of the African people for democratic rights is to be successful they must make a realistic assessment of the situation in Zimbabwe. African workers and peasants continue to search for a solution which is in their interests. Such a solution exists but as a precondition it needs the rejection of the nationalist movement's leadership and its political programme. It is not sufficient merely to have made changes in personnel; this solves nothing. In effect, this actually helps to split the liberation movement into those who benefit

and the African masses . . . Contrary to the claims of
the Mugabe government, it is abundantly clear that no
transition to socialism is taking place. The struggle of
the African people for socialism has been abandoned
in favor of encouraging the development of capitalism
in Zimbabwe—the so-called national democratic revolution. Such a political strategy, while clearly benefit-
ing a small section of Africans is, at the same time,
inevitably being carried out at the expense of the African masses.[29]

Another postwar observer characterized ZANU as a "broad liberation front based in a multiclass alliance" with an "educated petty
bourgeoisie" leadership and a mass peasant base.[30] ZANU strength in
the rural provinces of the Shona tribes as well as among urban Shona
speakers in Salisbury and other small towns in the eastern provinces
of the country only emphasizes the complexities that belie a simplis-
tic class-based analysis. ZAPU can be similarly characterized, except
that its mass base was grounded in urbanized workers, particularly in
the townships and cities of the Matabeleland provinces. The leadership
of both ZAPU and ZANU appears to have been directive rather than
consensual in nature; the occasional fractiousness of these movements
signifies not merely a lack of positive control but also the inability to
resolve differences among leadership elites.

After UDI in 1965, the British government imposed economic
sanctions, including a ban on importation of tobacco and sugar from
Rhodesia and a denial of access to capital markets. These actions were
followed by United Nations (UN) sanctions, to include measures to
deny the export of arms and oil to Rhodesia. The list of embargoed
goods was later expanded to include the products of Rhodesia's mines,
such as iron ore and chrome.[31] While Rhodesia became a pariah state,
such measures reflected only passive support for the African nationalist cause among Western governments. By contrast, active support for
ZAPU and ZANU came from a variety of Eastern Bloc governments
as well as nongovernmental organizations in Western countries. OAU
nations provided a combination of sanctuary and material resources
throughout the conflict. Zambia and later Angola and Botswana provided sanctuary for ZAPU base camps. ZANU was largely based in Tanzania and later Mozambique, with intermittent support from Zambia.[32]

While all parties to the Rhodesian conflict exercised a degree of self-restraint as the level of violence expanded, in general conflict between the Rhodesian government and the respective African nationalist movements was largely unrestricted and devoid of tacit cooperation between the insurgents and the government (e.g. by leaving control over some territory or population uncontested). While the black African majority offered a degree of passive support for the insurgents, the Rhodesian government did its utmost to crush the resistance, wherever and whenever it could be found, both within and outside Rhodesia's boundaries. The Rhodesian government also exploited tribal tensions to pit the resistance movements against each other as much as possible.[33]

The initial successes of the Rhodesian government's counterinsurgency strategy strongly limited the ability of ZAPU and ZANU to operate or move freely within the nation's boundaries prior to 1972. This problem was intensified when the government established a cordon sanitaire along the Rhodesian border to impede access into the country. As the war progressed, the Rhodesian Security Forces (RSF) conducted an aggressive series of cross border operations against ZANU and ZAPU bases. Because of these measures, both movements may be characterized as displaced insurgencies with relatively limited capacity to operate from domestic bases for much of the conflict. Although ZANU gradually succeeded in infiltrating the Mashonaland countryside, both groups operated mostly from an expanding web of sanctuaries in the Front Line States of southern Africa, using coercion to prevent the local population from reporting their activities, and setting up temporary base camps to support themselves while in-country.[34]

As will be discussed on more detail later in this chapter, ZANU and ZAPU employed differing strategies to overcome these obstacles. After a number of well-publicized failures, ZAPU operatives pursued clandestine operations in Ndebele-speaking areas, operating in small, semiautonomous cells.[35] ZANU, on the other hand, focused on eliciting popular support through a series of nocturnal outdoor meetings called *pungwes*.[36] While the ZANU approach corresponded to Maoist concepts of people's war, ZAPU adhered to insurgency methods that have been mischaracterized as conventional; the ZAPU strategic concept, capabilities, and practices were in fact based on a combination of Russian, Cuban, and Vietnamese methods.[37]

The origins of the ZAPU/ZANU split centered on the strengths and weaknesses of Joshua Nkomo's leadership, particularly with respect to his willingness to accept a compromise in the 1961 constitutional convention negotiations. Although both movements eventually adopted insurgent strategies designed to achieve political power emerging "from the barrel of a gun,"[38] they maintained effective public components, and ZAPU's leaders never lost their flair for clandestine operations combined with skillful diplomacy. ZAPU worked closely with the Soviets up to and during the Lancaster House Talks.[39]

Both ZANU and ZAPU blended African nationalist themes and varying forms of Marxist ideology in their strategic narratives. Both movements solemnized the memory of the Matabele and Shona uprisings of 1896–1897, emulating the spirit mediums who incited these revolts. ZANU adapted the Shona term *chimurenga* (struggle) to characterize the conflict against the Europeans, dubbing the nineteenth-century uprising the First Chimurenga; the Rhodesian insurgency became the Second Chimurenga. The ZAPU leader, Joshua Nkomo, was nicknamed the *chibwe chitedza* (slippery rock), as a reference to his relationship with the mystical Zimbabwean *ilitshe* (a divine rock) as well as his ability to escape arrest.[40]

Despite their use of African tradition as part of their ideologies, ZAPU and ZANU gravitated toward Marxist ideology as a means of mobilizing and indoctrinating the resistance. Key ZANU leaders, trained in China, actively and enthusiastically embraced the Maoist doctrine of people's war. These teachings were embodied in the ZANU *Mwenje One* manual, which included a recitation of the "National Grievances."[41] Similarly, ZAPU enunciated a six-point policy platform consisting of the following points:

1. "To fight for the immediate and total liquidation of imperialism and colonialism, direct and indirect; and to cooperate with any intentional forces as are engaged in this struggle;

2. To establish a democratic state with a government based on one man, one vote;

3. To foster the spirit of Pan-Africanism in Zimbabwe and the maintenance of firm links with Pan-African movements all over Africa;

4. To maintain peaceful and friendly relations with such nations as are friendly and peaceful towards us;

5. To eliminate the economic exploitation of our people; and

6. To foster the best values in African culture and thereby develop the basis of [a] desirable social order."[42]

ZANU and ZAPU relentlessly pursued these strategic goals throughout the conflict, linking these goals to the historical, socioeconomic, and political grievances and aspirations of the black African majority in Rhodesia. Both movements used their respective ideological frameworks as a justification for the use of violence, including acts of sabotage, terrorism, and civil war against the white Rhodesian regime and its supporters, both black and white.

At the same time, both the Soviets and the Chinese viewed the struggle of black African nationalist movements in Southern Africa, including but not limited to those in Rhodesia, as a logical extension of the international class struggle, a long-standing conflict between the urban and rural proletariat and the forces of imperialism evoked by Lenin and Mao Tse-tung.[43] As previously noted, in 1961 the Soviet government decided to use the KGB in a worldwide campaign to support national liberation movements against Western interests. At the same time, Chinese Foreign Minister Chou En Lai declared that Africa was "ripe for revolution." While the Soviets initially made inroads in Algeria, Ghana, and Egypt, China's strongest African partner was Tanzania.[44] These African states, together with Zambia (and eventually Mozambique), provided the strongest support in terms of finances, training, and sanctuary to black African resistance movements in Rhodesia. This support reflects the movements' often-meager resources as well as an increasing volume of support from Communist nations.

Strategies and Supporting Narratives

As previously noted, the strategies ZAPU and ZANU pursued diverged and evolved over the course of the insurgency. The ZAPU/ZANU split emerged largely because of ZAPU's emphasis, under Nkomo's leadership, on use of the Zimbabwean nationalist public component to elicit external support and to seek a negotiated settlement. ZANU's leaders argued that the political process had been tried and had failed and that a more radical approach, which was eventually based on the mobilization of the Zimbabwean black majority, was nec-essary. As noted previously, failure to achieve their objectives within the political process, combined with the Rhodesian government's suc-cessive banning of African nationalist movements, led both organiza-tions to expand the use of violence beyond the confines of the black townships. Both groups used mass media campaigns and direct recruit-ment to win popular support. While ZANU gained a strong support base among rural Shona speakers, ZAPU used its clandestine network to build support among urban workers, particularly in Ndebele-speak-ing areas. Although both movements failed to achieve demonstrable military successes throughout the balance of the conflict, ZAPU and ZANU propaganda emphasized the heroism of their fighters.

Both groups operated from sanctuary throughout the conflict, only contesting control over the Rhodesian countryside in the last two years of the war. Rhodesian geography was not especially favorable to guerrilla operations, offering little overhead cover or concealment. As the conflict wore on, the Rhodesian government employed a variety of measures to deny access to the insurgents, including (as already noted) establishing a cordon sanitaire along the Rhodesian border with Zam-bia and Mozambique and creating protected villages into which rural Africans were segregated. Although such measures hampered the insurgent movements' ability to operate freely within Rhodesian ter-ritory, they were unsuccessful in preventing the insurgents from per-forming clandestine operations to subvert the government's authority within the African majority.[45]

As previously discussed, the increasing ideological alignment of ZAPU and ZANU with their respective Communist supporters, together with OAU policies, shaped these movements' strategies. Both groups sought the elimination of white minority rule within Rhodesia, together with Zimbabwean independence, on the basis of black major-ity rule. The Western powers, specifically the United Kingdom and the United States, sought to minimize the impact of these demands, while agreeing to black majority rule and Zimbabwean independence in prin-ciple. With the onset of the Carter administration in the United States and the Thatcher government in the United Kingdom, the United States became a more vocal advocate of black African nationalist inter-ests across Southern Africa, while the British government adopted a more conservative stance.[46] The public components of both ZAPU and ZANU appealed to nongovernmental organizations in the West, particularly church organizations opposed to the policies of the Rhodesian and South African governments. Transnational connections between ZAPU and ZANU leaders and these organizations helped the movements elicit support from Western activists disposed to raising funds and providing nonmilitary support on behalf of these groups.[47]

The insurgents' strategic narratives were designed to appeal to an international audience and to the black African majority in Rhodesia. Neither the leading powers of the East nor West were particularly sympathetic to the plight of the Rhodesian government. The Communists portrayed Rhodesia as a capitalist puppet and the black nationalists as the repressed proletariat fighting for Socialist revolutionary ideals. State presses in Eastern Europe, Russia, and China heavily covered the African revolutionaries in a variety of languages and distribution locations. On the other hand, the positions of Western government and nongovernmental actors varied from studied indifference to active hostility to advocating continued white Rhodesian rule. In Britain, public sentiment was strongly against the legacy of colonialism, with the exception of a handful of conservative publications such as the London *Daily Telegraph*.[48]

The strategic narratives of ZAPU and ZANU emphasized the practical consequences of continued white minority rule as well as the legacy of African culture and injustices resulting from the Rhodesian government's policies. Both ZAPU and ZANU officially de-emphasized tribalism in their strategic narratives, preferring to describe the conflict in terms of white racism. As one source describes it, their argument was based on a "gut-level conviction that the white minority government had no right to be ruling them."[49] ZANU's strategic narrative was particularly well developed, taking the form of talking points to be used by ZANU cadres when addressing their black African audience. Examples of these themes include the following:

1. "Speak politely to the masses and to each other."
2. "No harassment."
3. "No strict speaking or beating."
4. "Deal with ZANU."
5. "We are our own liberator."
6. "Denounce in strongest terms [the] Internal Settlement."[50]

The insurgents' narratives likewise sought to invoke and preserve black Africans' historical memory of resistance to European settlers. In the ZANU hagiography, the Battle of Sinoia (April 28, 1966), arguably the first action of the Second Chimurenga, figures prominently. During this conflict, seven of twenty-one ZANU guerrillas were killed and were thereafter proclaimed heroes.[51] Another example of the invocation of Zimbabwean history was the ZANU Zimbabwe detachment, so named because it operated in and around the Zimbabwe ruins near Fort Victoria.[52]

The role of religion in the insurgent groups' strategic narrative revealed the complexities of the black Africans' past. Although some insurgents appear to have wholeheartedly embraced Marxist-Leninist ideology, their messaging also embraced Christian and traditional African religious themes. For example, the 1972 ZANU political program declared that the land belonged by right to black Africans "by almighty God."[53] Likewise, as the conflict continued, the insurgents' increasingly referenced spirit mediums in their messaging. For example, ZANU claimed that Zimbabwe African National Liberation Army (ZANLA) infiltration from Tete Province into northeastern Rhodesia was supported by a spirit medium named Chipfne, who also helped the guerrillas make contact with other spirit mediums, including one claiming to be a reincarnation of Nehanda, the spirit medium who incited the First Chimurenga in 1896–1897.[54]

Both ZAPU and ZANU used radio broadcasts to disseminate their messages. Efforts to unite their strategic messaging under the Patriotic Front (PF) failed, and ZAPU continued to transmit messages with Soviet support after the 1980 elections, with its characteristic call sign consisting of a lion's roar followed by gunfire and the song "ZIPRA Is Invincible."[55]

Structure and Dynamics of the Resistance Movement

Leadership

With few exceptions, the political leaders of ZAPU and ZANU were born between 1915 and 1930, whereas the military leaders were younger, having been born between 1935 and 1950. A key formative experience for many of these leaders was their education in various mission schools. The great majority of these men were at least nominally

Christian, their parents having been converted after the consolidation of British rule over its central African colonies. In a sample of thirty leading ZAPU, ZANU, and UANC leaders, 10 percent had at least a primary education, 33 percent had a secondary education, and 40 percent had a post-secondary education. Most had worked in white-collar jobs before joining the insurgency, with the largest number having worked as teachers (30 percent), lawyers (13 percent), and trade unionists (13 percent). As such, this group represented the elite of black African society in colonial and postcolonial Southern Rhodesia. Shona speakers outnumbered Ndebele speakers two to one, with tribal affiliations largely representative of the country at large. Of the seventeen black African nationalist leaders in Rhodesia with known religious affiliations, Methodists formed a substantial minority (27 percent), followed by Anglicans (13 percent) and Roman Catholics (10 percent).[56]

Key Leaders

Joshua Nkomo was born in 1917 in Semokwe. His education included three years at a secondary school, Adams College, and a baccalaureate degree from the Jan Hofmeyer School of Social Science (both in South Africa). A Kalanga, Nkomo held a number of jobs, including positions as a trade unionist, businessman, and Methodist lay preacher, before becoming president of the newly constituted SRANC in 1957. After the banning of the SRANC, Nkomo formed the NDP in 1960 and participated in negotiations with the British government over the provisions of the 1961 CAF constitution. Nkomo was criticized for agreeing to a split roll with legislative seats reserved for whites and blacks in the proposed constitution, but he was overruled by the NDP's governing council. Nkomo continued to lead ZAPU, the NDP's successor organization, through its dissolution in 1987. He was incarcerated from 1964 to 1974 and subsequently operated outside of Rhodesia until the conclusion of the Lancaster House talks in 1978. Throughout his career, Nkomo's stock in trade was the strength of his external partnerships, with both the East and the West as well as among the nonaligned nations of Africa and the Third World. In spite of his insurgent activities, Nkomo was regarded as relatively moderate and willing to pursue a negotiated settlement.[57]

Robert Mugabe was born in 1924 at Kutama Jesuit Mission where he grew up with his cousin James Chikerema. Mugabe is a Zezuru Shona speaker and was trained as a teacher at Kutama College before attending the University of Fort Hare in South Africa. A highly complex man, Mugabe demonstrated strong intellectual gifts early in life and rates as one of the most ideologically inclined of the black African nationalist leaders in Rhodesia. Mugabe's contemporaries at Fort Hare included Julius Nyerere (Tanzania), Herbert Chitepo (Rhodesia), and Kenneth Kaunda (Zambia). After teaching in Northern Rhodesia and Ghana, Mugabe joined the NDP as publicity secretary in 1961, continuing in this role with ZAPU until the ZAPU/ZANU split in 1963. Mugabe was incarcerated along with other black African nationalist leaders in 1964. While in confinement, he furthered his education through correspondence courses, earning degrees in law, economics, and administration from the University of London. Released in 1974, Mugabe rose to lead ZANU, representing the organization at the 1976 peace talks in Geneva. With the dissolution of the PF, Mugabe led the ZANU faction, ZANU-PF, which gained power in the 1980 elections and has continued to hold power in Zimbabwe for more than thirty-five years.[58]

Ndabaningi Sithole was born in 1920 in Nyamandhlovu. The son of a carpenter from the Ndau Shona tribe, Sithole attended a series of mission schools, graduating from the Dadaya Mission School in 1939. After teaching at the primary school level for nine years, he began studying for the Methodist ministry, joining the Methodist Church in 1950. Sithole studied at the Andover Newton Theological School in Newton, Massachusetts, from 1955 to 1958. Previously affiliated with the federationist Central African Party, he joined the NDP in 1960, becoming party treasurer. When ZAPU was formed, Sithole was appointed national chairman. He broke with ZAPU in 1963, becoming the ZANU party president until being ousted from power in 1975. As with other black African nationalist leaders in Rhodesia, Sithole was jailed in 1964 and released in 1974. While he was confined, Sithole completed and expanded his first book, *African Nationalism*, with an additional ten chapters and wrote a novel entitled *The Polygamist* and a historical work, *Obed Mutezo—The Story of an African Nationalist (Christian) Martyr*. After Mugabe rose to power in ZANU-PF, Sithole led the ZANU-Ndonga faction, entering into an alliance with James Chikerema and Abel Muzorewa from 1978 to 1980. Cary and Mitchell describe Sithole as "the great enigma of the nationalist movement," with an "unwavering determination in the pursuit of power," "a strong penchant for violence," and a "capacity for intrigue and political maneuver" that makes "even his political colleagues nervous."[59]

Jason Moyo was born in 1927 in Plumtree. An Ndebele speaker from the Kalanga tribe, Moyo achieved a primary school education and worked as a carpenter. Moyo was active in the SRANC, rising to the presidency of the Bulawayo branch in 1957. Moyo was detained twice for short periods in 1959 and 1960. He became a member of the NDP National Executive and rose to become treasurer and later vice president of ZAPU until his assassination in 1977. Operating from Zambia to escape incarceration, Moyo remained loyal to Nkomo throughout his time in party leadership. He shaped ZAPU's strategy during the period from 1964 to 1974 and continued to play an important role in oversight of ZAPU military operations until his death from a parcel bomb in Lusaka in 1977.[60]

Herbert Chitepo was born in 1923 in Watsomba, Inyanga. A Shona speaker, Chitepo attended Anglican mission schools before attending Adams College and Fort Hare in South Africa. After graduating with a bachelor of arts from Fort Hare, Chitepo worked as a Shona-language research assistant at the London School of Oriental and African Studies; while in London, he studied law at King's College and the Inns of Court, entering the bar in 1954. Chitepo practiced law in Salisbury from 1954 to 1959, joining the NDP in 1960 as a member of its National Executive. He left Rhodesia in 1962 to become the director of public prosecutions in Tanzania. The next year, Chitepo broke with ZAPU and was appointed ZANU national chairman. As the senior ZANU official to escape incarceration, Chitepo oversaw ZANU incipient guerrilla operations until he was assassinated in Lusaka, Zambia, on March 18, 1975. Although the Zambian government alleged that ZANU dissidents were responsible for his death, subsequent studies have asserted that he died as a result of Rhodesian direct action.[61]

James Chikerema was born in 1925 at the Kutama Jesuit Mission. A Zezuru Shona, Chikerema's father was the first teacher at the mission school and a practicing Roman Catholic. Chikerema achieved a secondary school education at St. Francis School before he moved to Capetown, South Africa. He was politically active early in his life, forming the Central African Social Club and demonstrating against the proposed CAF in 1953. Deported from South Africa in that year, he worked as a clerk and an insurance salesman in and around Salisbury. Together with three other black African nationalists, he formed the City Youth League (also known as the African National Youth League) in 1956, organizing a bus boycott in August of that year. When the City

Youth League merged with the SRANC in 1957, Chikerema became vice president, a position he later held in ZAPU. Chikerema was jailed in 1959 and released in 1963, subsequently leaving the country. He held the position of acting president of ZAPU, operating from Zambia until he broke with the ZAPU leadership in 1971 and formed the Front Line for the Liberation of Zimbabwe (FROLIZI). Allying himself with Muzorewa's UANC organization to oppose the new constitution proposed by the Rhodesian government, he eventually lost favor with the insurgent movements and was expelled from Zambia after Chitepo's death in 1975. Chikerema served as a delegate to the 1976 Geneva talks as a representative of Muzorewa's UANC organization. Describing himself as a "democratic national socialist," Chikerema abandoned the Roman Catholic faith, reverting to traditional African Shona religious beliefs and practice. Like Sithole, Chikerema was a complex, conflicted, and, in the end, unsuccessful leader.[62]

Lookout Masuku was born in 1940. Not much is known about his background, other than his Ndebele affiliation, or the specifics of his path to leadership within the ZAPU military structure, in part because of his continuous use of pseudonyms while active in the insurgency. He was the commander of ZIPRA through the conclusion of the 1979 Lancaster House talks and was subsequently jailed by the Zimbabwean government for conspiracy. Masuku contracted cryptococcal meningitis and died in 1986. He was interviewed by *Time* magazine in 1980, stating, "I behaved like any other youth . . . We wanted to vote and to be able to choose our own destiny. Instead parties were banned, people were arrested and killed, and there was nothing left but to wage an armed struggle." Although Masuku admitted to having killed informers, he denied responsibility for killings resulting from "lawlessness, banditry and blackmail."[63]

Josiah Tongogara was born in 1938 in Selukwe. He attended an Anglican mission school and later studied bookkeeping. He worked in Northern Rhodesia before joining the resistance. Tongogara was trained in China in 1966 and became a committed Marxist-Leninist. He rose to become the military commander of ZANLA in 1972. He was detained by the Zambian government in 1975 and charged with the murder of Herbert Chitepo. Acquitted and released in October 1976, he attended the Geneva conference and continued in his role as ZANU chief of defense. He was appointed commander-in-chief of the Zimbabwe People's Army (ZIPA) when ZANU and ZAPU sought to unify their military command structure. Tongogara was killed in an automobile accident just after the conclusion of the Lancaster House talks. A dynamic and charismatic leader, Tongogara was well respected by his subordinates and considered a rival to Mugabe in the postwar political lineup.[64]

Dumiso Dabengwa was born in 1939. Active in the City Youth League, he was incarcerated by Southern Rhodesian authorities from 1960 to 1963. Trained in the Soviet Union, he rose to become the head of ZAPU intelligence for the balance of the Rhodesian conflict. His ties to the Soviets were so close that his white Rhodesian opponents dubbed him the "Black Russian." Circumspect, rational, and soft-spoken, Dabengwa was instrumental in formulating and executing ZAPU operations through the end of the war. He participated in the Lancaster House talks and was later incarcerated by the Mugabe government, together with Lookout Masuku, for conspiring against the ZANU-PF regime. With the merger of ZANU and ZAPU in 1987, he was freed, subsequently holding Zimbabwean governmental positions, and has led a revivified ZAPU opposition party since August 2010.[65,c]

Solomon Mujuru (Rex Nhongo) was born in Enkeldoorn in 1949. He attended the mission schools in Kwenda and Rufaro at the primary level, as well as the Zimuto Secondary School. He then joined the ZAPU Youth League and was imprisoned from 1966 to 1968. A Zeruzu Shona speaker, he fought with ZAPU until the Chikerema/FROLIZI schism and then joined ZANU. Adopting the name Rex Nhongo as his nom de guerre, he underwent training in the Soviet Union, Egypt, Tanzania, and Bulgaria. Nhongo led missions into northwest Rhodesia from Mozambique and later took command of ZANLA while Josiah Tongogara was incarcerated in Zambia. In this capacity, he was designated commander of ZIPA during the attempt to unify the ZANLA and ZIPRA command structures. He was also present during the 1979 Rhodesian raid on the ZANLA base at New Chimoio. As deputy commander of the ZANLA forces, he succeeded Tongogara and served as commander of the Zimbabwean military forces after the 1980 elections. Nhongo's sunny disposition and roguish smile reflected a strong sense of humor while concealing an aggressive and ruthless character.[67]

---

[c] Dabengwa and his colleague Richard Dube (Gedi Ndlovu) could well be the KBG agents NED and ARTUR, respectively. NED was both a KGB and a GRU asset. ARTUR broke contact with the KGB after returning from training in Simferopol.[66]

91

As previously stated, the leadership of ZAPU and ZANU represented the best-educated and most politically active elements of the black African majority in Rhodesia. Although often strongly motivated by Marxist-Leninist ideology, these leaders frequently displayed a pragmatic streak with a human touch. Invariably outgunned and outfought by the RSF until the end of the war, they demonstrated a strong resilience and determination to continue the struggle against white European rule in Rhodesia. Of note is that virtually all of the ZAPU and ZANU leaders had been imprisoned at one time or another. Additionally the political leaders were often fractious and divided, motivated by ambition and in some cases suspicion of each other. This divisiveness led to a proliferation of competing insurgent groups, a situation that was sometimes exploited by the Rhodesian government. Although these movements were influenced by external supporters, especially in the case of Soviet influence on ZAPU, the forces driving the insurgent networks were a combination of interpersonal, tribal, and linguistic relationships. The strongest personalities rose to the top, exerting influence within these organizations. As one observer noted,[68] the complexity of the moral and psychological outlook of these men parallels Kipling's scathing critique of Irish leaders during the Troubles of the nineteenth century, in his poem *Cleared*:

> They never told the ramping crowd to card a woman's hide,
>
> They never marked a man for death—what fault of theirs he died?—
>
> They only said 'intimidate', and talked and went away—
>
> By God, the boys that did the work were braver men than they!
>
> (Rudyard Kipling, "Cleared")

Organizational Structure

The insurgent groups' organizational structures evolved out of the black African nationalist political parties and trade union movements in Southern Rhodesia, with only minor modifications as successive organizations were banned. The purpose of these structures was to mobilize political support and organize and control the activities of insurgents according to the policies established by the leadership. As the Rhodesian insurgency turned to violence and adopted Marxist-Leninist ideology, these movements began to reflect the principles of democratic centralism, emphasizing "unity of action in the midst of the struggle."[69]

Interpersonal rivalries, combined with tribal and linguistic divisions, resulted in brittle organizational climates within both ZAPU and ZANU. The long-term incarceration of the political leaders of both organizations complicated their inner workings until 1975. Gradually, both organizations developed a military command structure with young leaders capable of planning, coordinating, and executing insurgent operations, albeit often unsuccessfully, with greater security and deeper reach into Rhodesian territory.

The Zimbabwe African People's Union

ZAPU was formed on December 17, 1961, after the NDP was banned. It functioned with a political party structure that encompassed a president, a vice president, a national chairman, a national security organization, and a council. This executive body resembled a government in-exile in form, with departments overseen by secretaries and ministers. The ZAPU president served as the commander in chief of its military forces, and key secretary-level departments included those for administration, finance, publicity and information, foreign affairs, education, health, and women's affairs, as well as a national commissariat.[70]

The military component of ZAPU was ZIPRA. To integrate the political party headquarters and the military component, ZAPU established the Revolutionary Command Council as a representative body encompassing the key leaders of the military wing. The War Council was the executive-level body that made decisions based on the Revolutionary Council's guidance and passed those decisions directly to the ZIPRA high command. In many cases key leaders served as both civilian and military leaders.

Zimbabwe People's Revolutionary Army

The ZIPRA military command consisted of a ZIPRA commander, a commissar, and a chief of staff. The ZIPRA headquarters was organized into ten staff sections, including operations, intelligence, training and recruiting, transportation, and personnel, as well as chiefs of artillery, engineering, and reconnaissance. The military element was divided into two components, the guerrillas proper and the conventional forces. The conventional portion comprised a brigade with subordinate echelons, similar to the Soviet army.[71]

ZAPU organized its operations into four zones. The first two zones covered Matabeleland North and South and extended into the interior from the borders with Zambia and Botswana. The third zone stretched from the Urungwe Tribal Trust Lands (TTL) in north-central Rhodesia to Karoi, while the fourth zone extended from Belingwe to Vila Salazar.[72]

The Zimbabwe African National Union

The Zimbabwe African National Union (ZANU) functioned under the leadership of a central committee comprising a president, a vice president, a secretary general, and a military commander. As with ZAPU, the president acted as commander in chief of ZANLA, with a war council, the Dare reChimurenga. Its headquarters comprised eleven governmental divisions, each headed by a secretary. The secretary-level departments were administration, defense, foreign affairs, publicity and information, finance, health, production, education, women, and transport/social welfare, as well as a commissariat. Similar to ZAPU, ZANU integrated its military and civilian leadership to ensure consistent policy execution and the military command's subordination to the party line. Political commissars played a key role in the indoctrination of the guerrilla forces, preparing them to interact and politicize the rural population and instilling a degree of discipline and loyalty to civilian leaders.[73]

Organizationally, the party maintained committees at different geographical levels. The most localized committee was the village committee, and various village committees comprised a branch committee. Various branch committees constituted a district committee, with a geographic scope that often coincided with the territory of a particular tribal trust land. Multiple district committees helped constitute a provincial committee, which sent delegates to the biannual conference that elected the Dare reChimurenga.[74]

Village committees consisted of a chairman, secretary, political commissar, and police official. The committees at higher levels were similarly structured, and the main tasks of each of these committees included the mobilization and politicization of the resident population. Notably, village committees were also responsible for coordinating activities with ZANLA and ensuring insurgents in their area had sufficient food and clothing.[75]

Zimbabwe African National Liberation Army

Similar to ZIPRA, ZANLA divided its operational area into war zones extending from provinces along the Zambian and Mozambican border with Rhodesia (see Figure 4-1). The Zambia-Zimbabwe Zone confronted the RSF along the Zambezi River—a well-patrolled area infested with informants. The Mozambique-Zimbabwe Zone, established in 1972, extended south and west from Mozambique and was divided into three provinces, Tete, Manica, and Gaza, corresponding to the locations of their bases in Mozambique. The Botswana-Zimbabwe Zone extended to the western border of Rhodesia but was never operational. Each war zone was divided into sectors that followed a variety of naming conventions. For example, while Tete Province sectors were named after spirit mediums and revolutionary heroes, Gaza Province sectors were enumerated from I to IV. Sectors were broken down into detachments (operational areas). A detachment consisted of one hundred to two hundred men who operated in ten- to fifteen-man sections.[d] In addition to its sectors, Gaza Province had two independent detachments, one designated to fight in and around the Zimbabwe ruins.[77]

Institutionally, at the provincial level, the ZANLA leadership consisted of a field operations commander, a political commissar, an intelligence officer, and a medical officer, while the key officials at the sector, detachment, and section levels included a commander, political commissar, and officers for security, logistics, and medicine.[e] Of note is that the provincial political commissar managed the activities of the political commissars at the sector level, in particular by giving instructions regarding political training and propaganda in their levels. Furthermore, the political commissar at the section level had the

---

[d] Pandya provided a higher figure for detachments by noting that a detachment consisted of 300–450 men.[76]

[e] Moorcraft and McLaughlin noted ZIPRA also featured a similar hierarchical structure consisting of commanders and commissars.[78]

important task of politicizing the masses by giving speeches at pungwes that highlighted various grievances and government exploitation.[79, f]

The ZANLA high command sat at the apex of the insurgent army and was led by the ZANLA chief of defense, with other members including the chief of operations, the chief field commander, the political commissar, the chief of logistics, the chief of intelligence, the chief camp commander, and the chief of training.[81] The high command decided on the number of detachments to be deployed in a particular sector. To inform this decision, the chief of operations and chief field commander provided information on the activities and needs of forces in a particular sector, while the chief of intelligence briefed the high command on the security situation. Additionally, the chief of training and chief camp commander provided updates on the availability of newly trained insurgents at various camps and the number of ZANLA personnel available for redeployment. However, decisions regarding target selection was decentralized and in fact made by officials at the section level following consultations with local inhabitants.[82]

It should be noted that ZAPU and ZANU made at least two attempts to unify their military command structures. The first attempt, in 1967, involved the appointment of a Joint Military Council. The second, in 1975, likewise involved a joint military command structure under ZIPA. Both attempts failed because of interpersonal rivalry, disagreements over strategy, and differing perceptions of the other's capacity for insurgent operations. Dabengwa alleges that ZIPA's breakup in Mozambique and Tanzania resulted in bloodshed, as ZIPRA fighters were attacked in ZANLA camps; those who escaped "returned to their original bases in Zambia."[83]

---

[f] Other officers played important roles. For instance, logistics officers ensured that insurgents in lower levels obtained war supplies and that the latter were hidden in safe places. Furthermore, the security officer was responsible for providing protection against ambushes and during river crossings, as well as guarding against betrayal.[80]

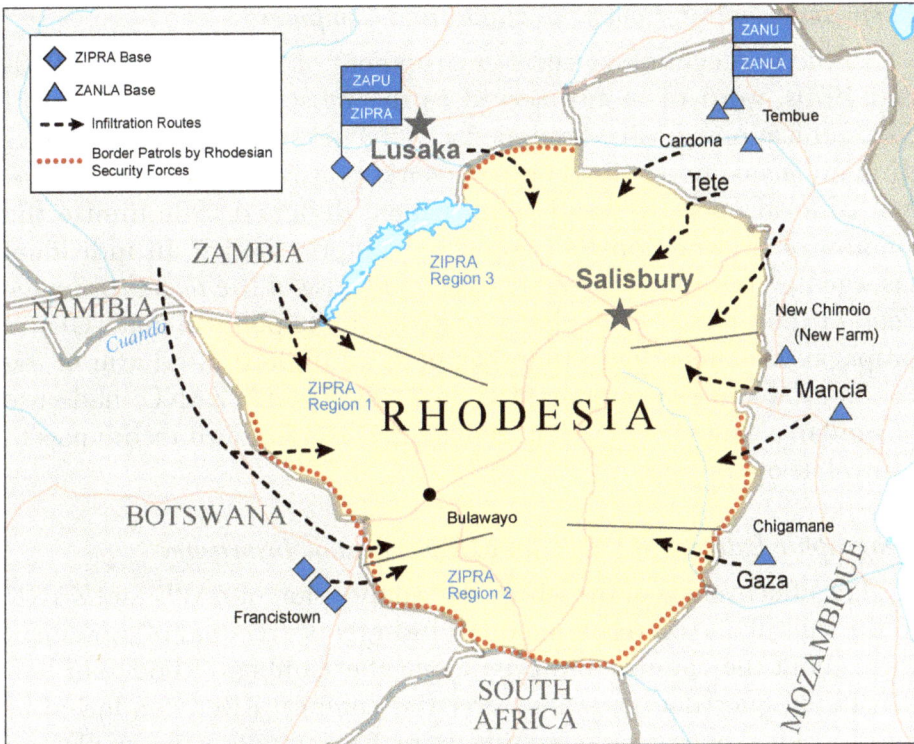

Figure 4-1. Insurgent staging and operational areas, 1974–1979.

Beginning in 1965, ZAPU operated a clandestine network consisting of two- to three-person cells that operated primarily in urban areas but later extended into the countryside. ZANU pursued a similar approach focused on operations in rural areas as its structure expanded in the mid to late 1970s. Initial efforts focused on reconnaissance, sabotage, and recruitment. Whereas the openness of Rhodesian geography and both natural and man-made obstacles impeded the insurgents' ability to penetrate the country and sustain themselves while there, the human terrain tended to favor the insurgents when tribal customs and language differences did not expose them to the Rhodesian government's counterinsurgency efforts. Throughout the hostilities, the insurgents operated in an environment in which the RSF owned the air and exploited this advantage to interdict guerrilla movements in the daytime. As the insurgents gained greater control over the countryside, the two groups extended their presence in Rhodesia, providing medical, administrative, and educational services to the population.[84]

Command, Control, Communications, and Computers

In the low-technology rural environment of the TTL in the 1960s and 1970s, word of mouth served as the most prevalent method of communications. ZAPU developed a "3-2-3 network" to relay messages through *mujiba* networks. These networks consisted of a "cut-out" system of local operatives and contact men, designed to minimize the amount of information that could be compromised if an individual insurgent were captured.[85] ZAPU also made use of the *Ibandla leZintandane*, or the Church of Orphans, to disseminate political material and propaganda. This organization began as an underground arm of the People's Caretaker Council (PCC), a ZAPU front.[86] ZANU made use of similar, although less well-articulated, word-of-mouth techniques to control its operations.

Geographic Extent of the Resistance Movement or Insurgency

The effectiveness of the Rhodesian suppression of ZAPU and ZANU in the mid-1960s severely limited the insurgents' freedom of movement throughout the country and even along the Zambian–Rhodesian border. Initial guerrilla operations over the Zambezi River, conducted by poorly trained operatives operating in small groups of section to platoon strength, met with failure. In 1973, the Rhodesian government closed the border with Zambia, forcing ZIPRA to seek new infiltration routes into Rhodesia from Botswana. Until the mid-1970s, the Rhodesian government was able to impede insurgent access through Zambia and, initially, Mozambique. This changed significantly when Portugal relinquished its African colonies and FRELIMO assumed power in Mozambique.

Whereas ZIPRA operated from bases in Zambia, Angola, and Botswana, ZANLA expanded its training base structure southward along the Rhodesian–Mozambican border. ZANU operated up to eighteen guerrilla bases and staging camps along the border between Rhodesia and Mozambique, with ZANLA headquarters located in Manica Province at New Chimoio and a training base located at Tembue in Tanzania. ZAPU operated eight guerrilla bases and staging camps in Zambia, plus two bases in Botswana and additional training facilities in Angola, with ZIPRA headquarters near Lusaka.[87]

There is much debate regarding the extent to which the respective insurgent groups were able to successfully penetrate the Rhodesian interior. A review of reported insurgent strengths derived from postwar sources indicates that insurgent strength inside Rhodesia grew from about 1,700 in 1976 to nearly 30,000 at the conclusion of the Lancaster House talks in December 1979. Initially, ZANLA used the difficult and densely vegetated terrain of the Mavuradonha Mountains to enable ingress from Tete Province into northeast Rhodesia (an area the Rhodesian defense forces dubbed Operation Hurricane). Although this region offered excellent prospects for insurgents to operate undetected, their strength in Operation Hurricane reached only 1,574 insurgents by January 1979. By contrast, insurgent strength opposite Manica Province (Operation Thrasher) grew from 400 insurgents in 1976 to 3,438 insurgents in January 1979; figures for Gaza Province (Operation Repulse) are 150 insurgents in 1976 to 3,548 insurgents at the beginning of 1979. ZANLA's main infiltration routes shifted southward, along the Mutirikwe and Mwenezi Rivers through the Gonarezhou game park as well as the Limpopo River along Rhodesia's southern border. Although the RSF mined the major transit routes along the border of Rhodesia and Mozambique, ZANLA insurgents were able to infiltrate from base camps in Manica Province through the eastern mountains near the Nyanga and Chimanimani National Parks.[88]

The insurgents' control of areas remained relatively modest through the end of the conflict, with ZANLA controlling four areas along the border with Mozambique, two in Hurricane, and one each in Thrasher and Repulse. For example, in the Takawira sector (Hurricane), ZANLA had established a provincial base with a network of underground tunnels; in the nearby Nehanda sector, insurgents sustained themselves, growing their own crops. At the same time, ZIPRA established a foothold in the Urungwe TTL near Karoi along the main road from Lusaka to Salisbury. ZIPRA operations extended south and west into North Matabeleland, contesting control over Sipolilo, Gokwe, and Tsholotsho in northern and western Rhodesia. By 1979, ZANLA and ZIPRA had extended their operations to the center of Rhodesia, including Salisbury, Bulawayo, and Gwelo.[89]

Although the ZIPRA front from Botswana was somewhat successful, the operational results of the efforts remained limited for a variety of reasons. Perhaps most important, the government of Botswana, while sympathetic to the nationalists' struggle, was unwilling to allow guerrillas to operate openly from its territory. This denied ZIPRA the ability to develop significant staging areas like ZANLA had in Mozambique. Additionally, the sparseness of the desert terrain, cross border raids by the RSF, and lack of strategic prioritization reduced the operational potential of the Botswana operation. Consequently, the estimated insurgent figures for Operation Tangent (encompassing North and South Matabeleland) reached only 884 by January 1979.[90]

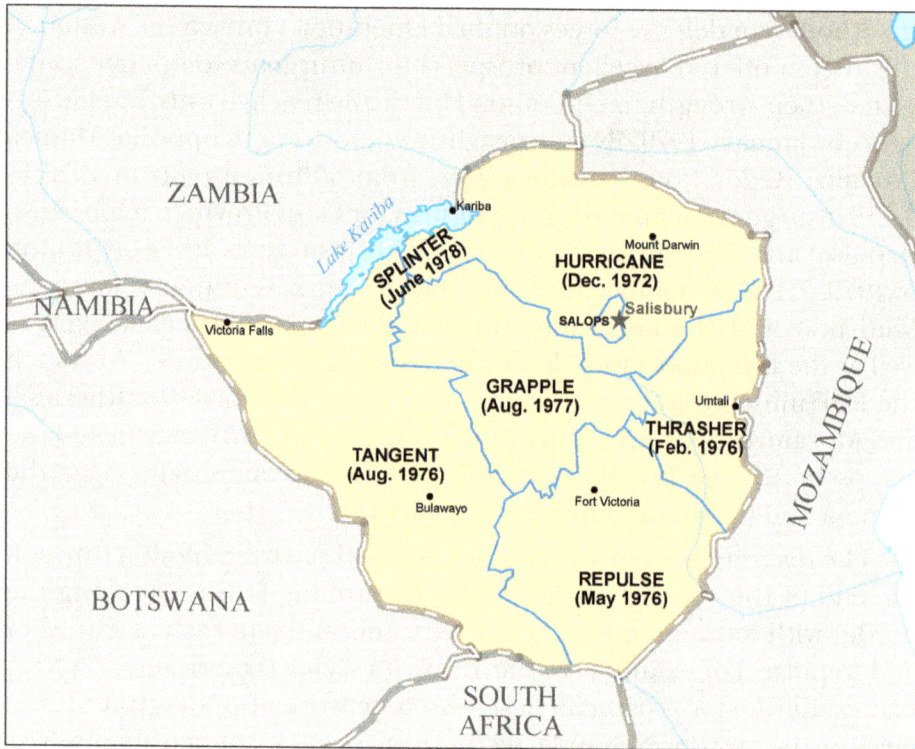

Figure 4-2. Rhodesian Security Forces operational zones.

The consequences of the insurgency's growth began to have severe effects on Rhodesia's governance, infrastructure, and economy from 1977 to 1979. As the conflict progressed, refugees streamed across the Rhodesian border into Botswana, Zambia, and Mozambique. In 1977, the refugee count in Mozambique amounted to 29,000, rising to 150,000 by the end of the war. By that time, almost 30,000 refugee children were attending schools in Mozambique. ZAPU established schools for refugee children in Zambia.[91] Overall the refuge count amounted to a quarter million by 1979. Transportation and medical services broke down. Cilliers notes that "rural bus services . . . had virtually collapsed by the end of 1978. . . . Of the thirteen Catholic mission doctors in

Rhodesia during 1975, only four remained by October 1978."[92] By the end of the war, nearly a half million children were out of school.[93]

Resources and External Support[g]

Both ZAPU and ZANU benefited from substantial external support throughout the conflict. This support took the form of direct and indirect funding, training, and equipment provided by state sponsors and nongovernmental organizations. The United States and Britain initially sponsored Nkomo, but as the Rhodesian insurgency turned violent, he sought and received support from Eastern Bloc nations. At the time of the ZAPU/ZANU split, the ZAPU account at Grindlays Bank in Dar es Salaam was frozen, with a mere £2900 in it.[94] The OAU established a Special Fund for Liberation, requesting each OAU member nation to donate 1 percent of their budget to this fund. The OAU allocated financial support based on the strength of insurgent cadres, in some cases encouraging the insurgent groups to impress recruits into service. Nigeria, Tanzania, and Algeria contributed to the fund and provided direct support.[95] In a 1974 meeting, Edward Ndhlovu, ZAPU deputy national secretary, stated, "At the moment we depend heavily on outside supplies of arms and funds to carry out our operations. . . . Logistic problems also limit recruitment and training . . . we don't have adequate arms and ammunition for most of those who want to join us inside."[96]

Among the African Front Line States, Zambia supported ZAPU, while Tanzania tended to support ZANU throughout the conflict. FRELIMO maintained contact with ZAPU through the Chikerema schism in 1970. Although FRELIMO initially favored ZAPU over ZANU, ZANU's willingness to fight alongside FRELIMO guerrillas in Mozambique worked to the advantage of both parties, with ZANU picking up operational experience along the way. As one of the ZANLA cadres involved in the effort, Samuel Mamutse (Urimbo) said, "Our mission was to go and be taught by FRELIMO, how they operated, how they stayed with the masses . . . the relationship between the party and military forces and how to mobilize the masses."[97]

---

[g] See chapter 5 for more detail regarding tactical support the Soviet Union and China provided to ZAPU and ZANU, respectively, and the role regional countries played in supporting both insurgent movements.

In a 1979 taped interview, ZANU officials, including Josiah Tongogara, claimed to have received £20,000 from a Christian aid group, promises of support from Oxfam, and clothing and medical supplies.[98] The World Council of Churches funded ZAPU and ZANU, as well as other insurgent movements in southern Africa, through the auspices of the Programme to Combat Racism. Although this support became controversial, the program gave a grant to the PF in 1978.[99] Similarly, Canon John Collins of St. Paul's Cathedral in London established an International Defence and Aid Fund (IDAF) to provide money for legal defense and support of families whose husbands had been incarcerated in Rhodesia and South Africa. Because the Rhodesian government routinely tried captured insurgents in criminal court, this support included the defense of imprisoned ZAPU and ZANU fighters.[100,h] Western groups also provided medical aid to ZANU through the auspices of the Zimbabwe Medical Aid organization.[102]

The Soviets began providing financial support to the nascent black African insurgency in Rhodesia in 1961, with a grant of $8,400 to the NDP. In 1962, Nkomo requested £150,000 in financial assistance. Soviet records show that ZAPU received $19,000 in 1963, $20,000 in 1965, and $28,000 in 1966. Initially, arms shipments to ZAPU proceeded through Dar es Salaam, with an intermediate stocking point at Mbeya on the route to Lusaka. Some of these shipments were diverted, so after Angola became independent from Portugal, the Soviets established a more reliable supply route through that country. In 1976, Nkomo requested a total of 6,750 small arms, including AK-47 rifles and SKS carbines and pistols, as well as rocket-propelled grenades, mortars, recoilless rifles, trucks, and river-crossing equipment. By the end of the war, ZIPRA was equipped with handheld SA-7 antiaircraft missiles, which were used to good effect during the last year of the war. The Soviets even supplied ZIPRA with armored vehicles, although these were never used in Rhodesia during the war.[103] ZANLA, supported by various OAU nations, principally Tanzania and Mozambique, as well as the Chinese, was less lavishly equipped than ZIPRA. The RSF identified this as a weakness and launched a series of cross border operations to strike ZANLA base camps and supply routes in Mozambique.[104]

---

[h] In 1982, after the conclusion of the Rhodesian Bush War, ZAPU applied for IDAF funding to defend the fourteen ZIPRA leaders, including Masuku and Dabengwa, against charges of treason. IDAF refused to honor this request because granting it would have constituted interference in Zimbabwean internal affairs.[101]

As previously stated, the Soviets and Chinese began providing training assistance to ZAPU in the early 1960s, with ZANU cadres going to China and Ghana after the two groups split. Some cadres were also trained in Cuba. The Chinese training program at the Nanking Academy in Beijing included political indoctrination, military strategy and tactics, and instructional techniques. As the ZANU cadres established training bases in Tanzania, Tanzanian People's Defense Forces instructors assisted them as drill and physical fitness instructors. In 1969, a group of eight Chinese instructors arrived in Tanzania to provide training in heavy weapons, reconnaissance, and sabotage; this cadre was later expanded to twenty instructors. Chinese training methods emphasized the psychological formation of ZANLA fighters through political indoctrination and drill. ZANLA also benefited from its relationship with FRELIMO starting in 1970. Operating with seasoned FRELIMO insurgents in Mozambique gave ZANLA operatives practical experience with field craft and clandestine operations.[105]

ZIPRA cadres attended training courses in a variety of Eastern Bloc and nonaligned countries, including the Soviet Union, China, Algeria, Cuba, Bulgaria, and North Korea. This training regime was more sophisticated and diverse than the training afforded to ZANLA, consisting of extensive instruction in clandestine operations, including sabotage and espionage, as well as ideological instruction and technical military training.[106] Shubin reports that ZAPU cadres attended the Institute of Social Sciences for political training. Beginning in July 1977, at Nkomo's request, Soviet and Cuban advisors began training ZIPRA fighters at base camps in Angola and Zambia. By this time, ZIPRA benefited from the arrival of more than five thousand recruits. The Soviet training regime consisted of a two-month program including small-unit tactics to company level as well as crossing water obstacles. Although the primary focus of this program was training in conventional operations, the Soviets also provided training in guerrilla tactics, "in case of temporary setbacks of the regular forces."[107] By July 1978, ten thousand ZIPRA fighters had been trained in Angola. ZIPRA also received training on armored vehicles in the Soviet Union. In 1978, the Soviets also sent a three-man advisory team to Lusaka to provide ZIPRA additional training and assistance in unconventional warfare tactics. In 1979, at least six Cuban instructors and one Soviet warrant officer were killed by the Rhodesian Air Force during an air strike on a ZIPRA training base in Angola.[108]

Political Activities

Although neither ZAPU nor ZANU adhered to a strategy of nonviolent resistance, their use of violence ebbed and flowed over the course of the conflict, with intermittent attempts to achieve a negotiated settlement from 1961 through the end of the war in 1979. The black African nationalist rejection of the 1961 Constitution led to a period of gestation culminating in the mass arrest of the black African national leaders in Southern Rhodesia in 1964. It was during this period that black-on-black violence increased, as the Southern Rhodesian government reverted to the control of the white supremacist RF, and the ZAPU and ZANU organizations took form. After the UDI, the die was cast, but because the British government refused to recognize a Rhodesian government that lacked black African support, the negotiating track remained open, if dormant.

Table 4-2 summarizes the evolution of constitutional proposals for Rhodesia/Zimbabwe from the failed 1961 negotiations on the constitution to the 1979 Lancaster House Agreement (not including the British–Rhodesian negotiations conducted in 1966 on the *HMS Tiger* and in 1968 on the *HMS Fearless*). All of these constitutional schemes included an A roll with property and education qualifications designed to exclude blacks, together with a less restrictive B roll. The 1976 negotiations featured a third roll; in the Rhodesian government proposal, this roll would be restricted to white voters, and in Nkomo's counterproposal it would be open to all voters.

Table 4-2. Proposed constitutional arrangements: 1961–1979.

Constitution/ Proposal	Legislative Structure	Voting Provisions
1961 and 1965 (UDI) Constitutions	Unicameral– 65 seats	A Roll–50 seats B Roll 15 seats
1969 Constitution/ 1971 internal settlement proposal	Bicameral– 66-seat House 23-seat Senate	**House–** A Roll–50 seats B Roll–16 seats **Senate–**10 seats reserved to whites, indirectly elected; 10 indirectly elected tribal chiefs; 5 Shona and 5 Ndebele; 3 senators selected by the head of state
1976 negotiations	Unicameral– Rhodesian proposal: 75 seats Nkomo proposal: 108+ seats	**Rhodesian proposal–** A Roll–25 seats B Roll–25 seats C Roll–25 seats **Nkomo proposal–** A Roll–36 seats B Roll–72 seats C Roll–up to 986 seats
1978 internal settlement	Bicameral– 100 seats	**House–** A Roll–28 seats B Roll–72 seats **Senate–**10 seats reserved to whites, indirectly elected; 10 unrestricted seats, indirectly elected; 10 indirectly elected by the Council of Chiefs
1979 Lancaster House Agreement	Bicameral– 100-seat House 40-seat Senate	**House–** A Roll–20 seats B Roll–80 seats **Senate–**10 seats reserved to whites, indirectly elected; 14 unrestricted seats, indirectly elected; 10 indirectly elected by the Council of Chiefs; 6 selected by the head of state

Sources: Gann and Henriksen, *Struggle for Zimbabwe*, 57, 136–139. See also *Southern Rhodesia: Report of the Constitutional Conference Held at Lancaster House.*

Under the Labour government of Harold Wilson, the British insisted that Rhodesia would not be recognized without the support of the black African majority. In 1971, after the British government had reverted to Tory rule, the Rhodesian government sought British recognition based on the provisions of the constitution adopted by the RF-led government in 1969. This led to the formation of the United African National Council (UANC) under the leadership of Bishop Abel Muzorewa, formed to block British acceptance of the 1969 Constitution. Subsequently, Muzorewa attempted to create a united front including ZANU and ZAPU. Although this effort fell apart, a ZANU splinter group under Ndabaningi Sithole renounced violence in 1975, while ZAPU and ZANU formed the PF under OAU pressure to conduct the peace negotiations held at Geneva in 1976 and at Lancaster House in 1979.[109]

Throughout the conflict, the "struggle for Zimbabwe" was accentuated by the escalation of violence punctuated by successive unsuccessful attempts at a political solution until the conflict finally ended, or at least was temporarily halted, at the negotiating table. The RF government's suppression of the black African nationalist organizations put an end to lesser forms of resistance, including strikes, boycotts, and incidental acts of sabotage such as the rash of petrol bombings and attacks on churches, schools, and communications facilities that took place in 1962.[110,i] Although the white-dominated Rhodesian government was relatively impervious to penetration by the insurgents, it became more and more dependent on black African support, particularly within the security forces and the police, as the conflict progressed.

After the Rhodesian government banned ZAPU and ZANU, neither group was able to operate freely within Rhodesian territory until the 1980 elections following the Lancaster House talks. As the record of that electoral contest demonstrates, the respective insurgent groups were well able to establish themselves as effective political parties capable of contesting the elections within their respective tribal bases.

Throughout the conflict, all participants increasingly used subversion and psychological operations with varying degrees of priority and success. In retrospect, the insurgents' strategic narratives appealed

---

[i] The Rhodesian government reported seventy-three petrol bombings from the beginning of 1962 through February 1963. Thirty of the bombings were targeted against individuals, and the remainder against property.[111]

more directly to the interests and values of the black African majority, despite the Rhodesian government's successive attempts to establish a successful counternarrative.[112] Although the RSF established a psychological operations unit in 1977, ZAPU relied on the aforementioned Church of Orphans to disseminate its messages.[113] As noted previously, ZANU tended to rely on direct contact through *pungwes* to get its message out to rural villagers, using small teams focused on political indoctrination. This approach emphasized a simple, clear, and consistent message. Despite the violence the insurgents exerted against the black African population, the lack of media among the rural population enabled ZANLA and ZIPRA forces to shape local perceptions of events. Atrocities committed by guerrillas were attributed to the government, and Rhodesian raids were portrayed as attacks on civilians and refugees. Poor education and lack of sophistication did not inhibit the success of the insurgents' information operations because the insurgents spoke the language of the people and understood what messages would work.[114]

As efforts matured, ZAPU and ZANU also made use of printed materials such as leaflets, posters, and newspapers, as well as radio broadcasts. The Voice of Zimbabwe was the primary ZANU station, and the Voice of the Revolution was ZAPU's. Most broadcasts were based in neighboring countries, such as Zambia, Tanzania, and, eventually, Mozambique, making it nearly impossible for the Rhodesian government to interdict guerrilla messaging. These radio stations were heavily supported by the Soviets and Chinese and often had foreign Communist technical advisors on hand. ZAPU's Voice of the Revolution was also afforded significant airtime on Radio Moscow.[115]

Both insurgent groups had limited capacity to provide services to the populace inside Rhodesia. Beginning in 1971, ZANLA began to focus on establishing a support base within the local tribal populations. As one observer describes it, "Particular attention was devoted to learning about local grievances and then offering practical solutions to these problems."[116] As the war progressed into the late 1970s, a flood of refugees poured over the borders into Mozambique, Zambia, and Botswana. As previously indicated, by early 1977, Rhodesian refugees living in Mozambique reached 29,000.[117] This number increased to 160,000 in camps run by FRELIMO and ZANU by 1979, with another 90,000 refugees in ZAPU camps in Zambia and Botswana. Neither ZANU nor ZAPU had the resources to support a displaced civilian population of this magnitude. The UN High Commission for Refugees stepped in, committing $3 million in aid. Although the UN agency made every attempt to ensure that its aid was not used to support the insurgency proper, the insurgent groups administered the aid to the refugees. Within the camps in Mozambique, ZANLA organized what limited medical and educational services it could provide, using doctors and teachers within its organization. Likewise, ZAPU established schools for refugee children in Zambia.[118] From a warfighting perspective, the sizable refugee population in neighboring countries was relevant because it served as a potential source of recruits.

Recruitment

In the early years of the conflict, both ZAPU and ZANU experienced difficulties in recruiting a sufficient number of males for military activity, and thus both relied on press-ganging, particularly because funds from the OAU were dispersed based on the number of recruits in training camps.[119] Indeed, ZAPU established a military wing in the mid-1960s, but as Brickhill noted, it was only after a decade of war that its personnel total exceeded one thousand.[120] By the end of the conflict, the organization had over twenty thousand insurgents, many of whom had been incorporated between 1976 and 1978.[121] The party played a fundamental role in recruitment. Sixty percent of recruits had been members of the ZAPU youth league, many of whom were recruited in urban settings where clandestine elements of the party remained intact. Many recruits had already been exposed to political activism through trade union activities, and over two-thirds of recruits came from families in which at least one parent was a member of illegal party units.[122]

Demographically, ZIPRA was a young fighting force, with 84 percent of members between eighteen and twenty-five, and 8 percent were over twenty-six.[123] More than half of recruits were employed before joining the insurgent force, and over 53 percent participated in urban wage employment.[124] Reportedly, 50 percent of recruits were from peasant backgrounds, yet Brickhill concluded that ZIPRA was noticeably urban and proletarian because many recruits had been exposed to urban capitalism, trade unionism, and political activity.[125]

Botswana played an important role in ZIPRA recruitment efforts. The number of Zimbabweans, many of whom were ZAPU members, who crossed into Botswana rose from 5,712 in 1976 to 25,300 in 1978. These entrants were accommodated at transit centers in Selibe Phikwe and Francistown, although some subsequently moved to a refugee camp at Dukwi.[126]

In the case of ZANLA, a member of the central committee, Maurice Nyagumbo, was responsible for recruiting individuals for military training outside of Rhodesia. His official title, secretary for public affairs, was a misnomer intended to mislead Rhodesian authorities of his true responsibilities.[127] ZANLA can perhaps be labeled as a peasant army because almost 90 percent of recruits were of peasant background, with the remainder consisting of urban residents.[128] ZANLA insurgents often engaged in recruitment once they entered Rhodesia from the Tete province in Mozambique.[129] Additionally, a large number of recruits were students and teachers at schools close to the border with Mozambique, such as St. Augustine's Mission, Old Umtali Teachers College, Mutambara, Sunnyside, Biriviri, Mount Selinda, Bonda, and Mary Mount.[130] Recruitment of these educated recruits often occurred at *pungwes*. As noted by Pandya, in the day after a *pungwe*, teachers and students often engaged in extensive political discussions at school, sometimes with ZANLA personnel present, who would remind them, among other grievances, of the inferior education the settler regime imposed on them. This was often enough to convince students and teachers to leave en masse for ZANLA camps in Mozambique. One school official at St. Benedicts Mission School noted that his best and brightest students crossed the border into Mozambique.[131]

Once, in Mozambique, local villagers put the border crossers in contact with ZANU and ZANLA officials, who would then direct recruits to camps in Manica Province in Mozambique that were established to screen recruits.[132, j]

---

j Some recruiting was conducting in urban settings, particularly Salisbury, Bulawayo, and Gwelo, and such efforts relied on existing social networks. Specifically, recruiters would typically talk to individuals at bars and social gatherings, and if the outreach was well received, the approached individual often spoke with friends and convinced them to head to Mozambique as well.[133]

Methods of Warfare

The Rhodesian insurgents targeted a combination of white settlers operating farms across the countryside, as well as black Africans who either supported the settlers or were considered government supporters. Although there were certainly instances of indiscriminate violence, particularly in the latter years of the war, even the targeting of white settlers was at times discriminate, focusing on farm owners with reputations for badly treating their black African workers. Likewise, the use of terror to coerce the black African population was not simply random; the insurgents and the government alike used violent methods to dissuade the population from providing support to the other side. Until 1972, only two insurgent attacks targeted white farmers and the level of black-on-black violence remained high, targeting tribal chiefs friendly to the government, black African civil servants and policemen, and relatively prosperous black Africans suspicious of the insurgents' promises. This targeting of pro-government black African leaders was so effective that two-thirds of the local councils were rendered inoperative. As Gann and Henriksen put it, "The nationalists persuaded the willing by stirring appeals, the hesitant by promises, the recalcitrant by terror."[134] In conjunction with attacks on civilian personnel, the insurgents targeted economically significant objectives, including livestock, farm buildings, and infrastructure. The insurgents made promiscuous use of land mines to disrupt the transportation of commercial goods throughout the countryside.[k] Only later in the war did this effort result in effective attacks on urban areas, oil storage facilities, and civilian aircraft.[136] Yet even in the later stages of the war, the insurgents were not able to effectively target government or military installations. Assassination attempts likewise proved unsuccessful.

Operationally, the insurgents sought to demonstrate the weakness of the Rhodesian government to force it to overreact to their provocations and to intimidate the white European population and any black Africans inclined to support the government. As the war progressed, the focus shifted to establishing effective control over the black African population, isolating urban areas, and eventually achieving military dominance. As operations progressed and contact with the local

---

[k] Wood noted that, from 1972 to 1980, land mines (mainly Soviet TM-46 mines) laid by ZANLA and ZIRPA resulted in 2,500 disabled vehicles and 5,000 casualties (including 632 deaths).[135]

population increased, the guerrillas were faced with elements of the population who were unsympathetic to their cause or sympathetic to the government. When they could not gain support from the local population willingly, ZIPRA and ZANLA often sought compliance by using terror and intimidation, such as torturing, mutilating, or assassinating suspected sympathizers. ZIPRA was generally more selective in its targeting, committing fewer atrocities than were attributed to ZANLA.[137]

Although intermittent peace negotiations occurred in the mid to late 1970s, both ZANU and ZAPU envisioned a culminating campaign to seize power. In the case of ZAPU, this took the form of a plan for a hybrid conventional offensive combined with the uprising of guerrilla cadres within Rhodesia, dubbed Operation Zero Hour.[138] The ZANU counterpart to Operation Zero Hour was the Goere reGukurahundi (The Year of the Storm), which sought to overwhelm the RSF by saturating the countryside with irregular ZANLA forces.[139] These military efforts were generally well synchronized with the insurgent movements' political strategies. Frequently, insurgent operations were timed to coincide with meetings of the OAU Liberation Committee, demonstrating results intended to buttress the argument for continued support.[140] All parties to the conflict continued to operate while negotiations were ongoing, seeking to influence the outcome of the negotiations.

In the early stage of the conflict, ZIPRA and ZANLA used relatively similar guerrilla tactics. After receiving training in Zambia, small groups, sometimes operating in platoon to company strength, would infiltrate across the Zambezi River to gain access to Rhodesian territory (see Figure 4-3). When moving in significant numbers, the insurgents were vulnerable to being detected by the RSF, whose advantages in mobility and firepower invariably resulted in the destruction of insurgent raiding parties. Without the ability to maintain a permanent presence inside Rhodesia, guerrilla units depended on bases in garrisons at training camps located in Tanzania and Zambia (and later Mozambique). From 1977, ZIPRA began a concerted effort to expand operations into Rhodesia from Botswana, primarily using it as a transit route to bases in Zambia and Angola.[141]

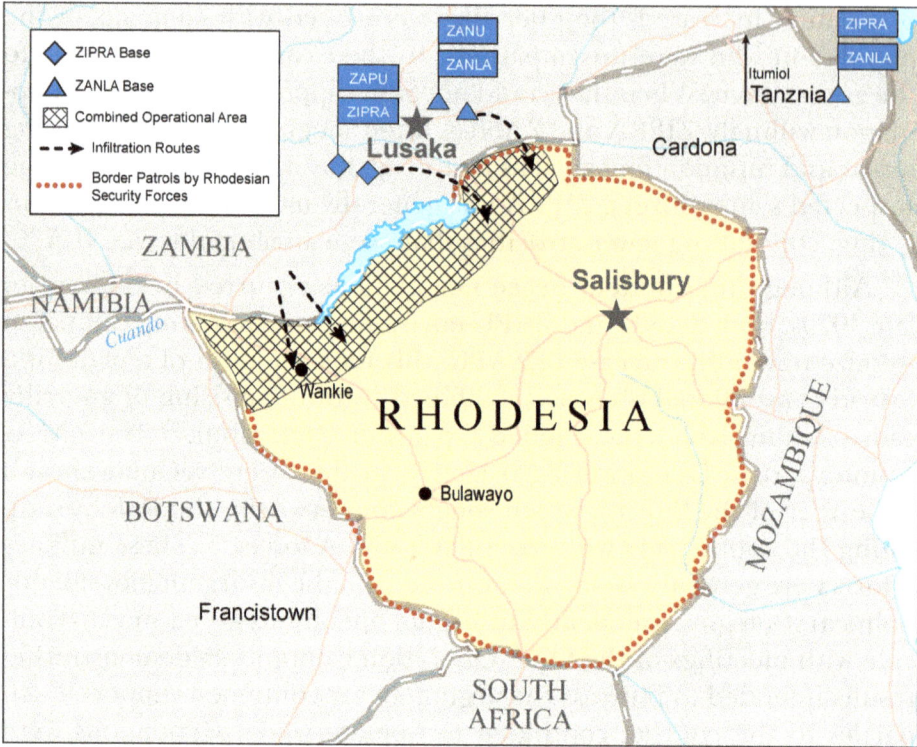

Figure 4-3. Initial insurgent incursions into Rhodesia.

Initially, these guerrilla patrols were not well supported by a secure clandestine infrastructure inside Rhodesia. Their levels of experience, training, and discipline were questionable, and they often lacked first-hand knowledge of the areas in which they were moving. Gradually, however, the insurgents gained greater freedom of maneuver inside Rhodesia. While the guerrillas could not control terrain, they could effectively deny control to the Rhodesians. With intelligence and logistical networks in place, the guerrillas were starting to tip the balance in their favor. In particular, ZANLA's improved training, coupled with combat experience gained during joint efforts with FRELIMO, made a significant difference. Increasingly guerrillas were able to effectively move long distances from campsites to target areas at night. After engagements, they were often able to disengage and effectively elude pursuing forces.[142]

When conducting raids and ambushes, small groups of insurgents typically moved from base camps at night. Although striking at sunrise offered the advantage of surprise, striking late in the afternoon offered secure egress in the hours of darkness. Upon making contact with the RSF, the insurgents typically attempted to withdraw,[1] but there are some notable exceptions, both early and late in the war, when they chose to stand and fight.[144] For example, during the battle of Wankie in 1967, a combined force of ZIPRA and South African ANC guerrillas stood and fought against the Rhodesian African Rifles, resulting in substantial casualties on both sides.[145] Both ZANLA and ZIPRA stood and fought when the RSF conducted cross border attacks on their base camps, such as the 1979 assault on the ZANLA base at New Chimoio and an attack on a ZIPRA regular battalion encamped in Zambia that same year.[146]

Early in the conflict, the insurgents established a pattern of sabotage attacks on economic and infrastructural targets. They routinely deployed land mines, resulting in Rhodesian countermeasures that resembled the later American response to improvised explosive devices in Iraq and Afghanistan, including the introduction of armored trucks with *v*-shaped chassis.[147] These actions were consistent with the ZANU Five-Point Plan adopted in May 1964, which called for the following:

1. Targeting Rhodesia's main roads and bridges

2. Erecting roadblocks to obstruct RSF movement

3. Destroying livestock and crops on European farms

4. Attacking power and communications infrastructure, including telephone lines and electric pylons

5. Raiding government institutions, including native commissioners' offices, police stations, and white-owned shops in African townships

Indeed, for the most part, the early cross border raids of the 1960s, as well as incidents perpetrated by in-country insurgents, focused on these kinds of targets. For example, one group of saboteurs, including Emmerson Mnangagwa (who is now the current president of Zimbabwe), Mathias Maloba, and Jimmy Munyavanhu, attacked the railway

---

[1] Moorcraft and McLaughlin noted that a "willingness to accept a few casualties might have shortened the war considerably. Sometimes remote farmhouses defended by a single family would hold out against 20 or 30 guerrillas equipped with mortars and rockets. Although this ultimately did not matter, in that the guerrillas achieved their war aims, their irresolute tactics prolonged the conflict by years. As one Rhodesian officer commented in 1979: 'If we had been fighting the Viet Cong, we would have lost the war a long time ago.' "[143]

line in Victoria (Masvingo) Province, blowing up a train carrying goods before being captured by the Rhodesian Sabotage Squad. In similar fashion were the operations of the ZANU Crocodile Gang, a five-man insurgent team (of which Mnangagwa was a member) responsible for the first black-on-white assassination (of P. J. Oberholtzer) and the failed attack on the Nyanyadzi Police Camp on July 1, 1964. This group also conducted a roadblock on the Chikwizi Bridge the following night, mistakenly assaulting a black African traveler.[148] The objective of the seven-man ZANLA guerrilla team that engaged the RSF at Sinoia on April 28, 1966, was to attack white farmsteads and cut power lines.[149] A third group's activities culminated in the killing of J. H. Viljoen on May 18, 1966.[150]

The insurgents escalated their tactics as the conflict wore on into the 1970s. The culmination of insurgent sabotage attacks took place on December 11, 1978, when an oil tank farm in Salisbury was attacked by rockets and sustained substantial damage.[151,m] This attack, as well as the two Air Rhodesia Viscount aircrafts ZIPRA guerrillas shot down with SA-7 missiles on September 3, 1978, and February 12, 1979, reflected the insurgents' increased sophistication and capacity for lethal attacks on nonmilitary targets.[153] Losses totaled 102 passengers and crew from these two attacks, including ten survivors killed on the ground by ZIPRA guerrillas.[154]

Many of these attacks were accompanied by violent threats designed to intimidate the population, especially white Europeans. For example, the Crocodile Gang left two notes after the Chikwizi Bridge incident, one of which read, "Ian Smith Beware. Crocodile Group on Confrontation. Political. White Man is Devil."[155] ZANU made at least one attempt to assassinate Ian Smith, but it was thwarted by the Rhodesian Special Branch.[156]

White missionaries represented perhaps the most controversial of the insurgents' targets. Missionaries in the late 1970s were especially vulnerable because they were dispersed throughout the countryside and predisposed to going unarmed at a time when even Rhodesian farm wives were carrying firearms. In July 1978, the Rhodesian Ministry of Information published a lengthy tract that cited twelve attacks on church and Red Cross aid workers from 1976 onward, resulting in

---

[m] As indicated by Independent Television News (ITN), both ZANU and ZAPU claimed responsibility for this attack.[152]

thirty-one deaths, two wounded, and one person missing and reported dead. The majority of these attacks were on members of Roman Catholic religious groups operating in Rhodesia, including priests, nuns, and a retired Catholic bishop. The deadliest of these attacks took place on a Pentecostal mission, the Elim mission at Vumba in the Rhodesian Eastern highlands, killing eight British missionaries and four children on June 23, 1978. Both ZAPU and ZANU were implicated in these various attacks, which extended from the border with Mozambique to the Bulawayo area.[157]

Other criminal acts the insurgents committed included impressing black Africans into serving with the guerrilla forces, including conscripting children. Examples include the raid on St. Albert's Mission school in July 1973, during which ZANLA guerrillas abducted 273 students, and the ZAPU raid on the Tegwani mission, during which more than 400 were kidnapped.[158] The insurgents were also capable of petty thievery. The propensity for engaging in criminal activity varied among insurgents and largely depended on how tolerant the local commander was to such activities rather than on the group's overall governing policy. ZIPRA was generally more disciplined and discriminating in the use of violence.[159] Neither group appears to have used extortion as a means of acquiring money or sustenance, nor did they exercise violence outside of Rhodesia proper, except in self-defense or occasional infighting.

Table 4-3. Insurgent weapons, munitions, and equipment

Nomenclature[a]	Description	Origin(s)
Small arms		
Tokarev (type 51)	7.62-mm self-loading pistol, 8-round magazine	Russia, China
AK-47 assault rifle (M-22)	7.62-mm single-shot and automatic rifle, 30-round magazine	Russia, China
SKS automatic carbine (type 56)	7.62-mm single-shot carbine rifle, 20-round magazine	Russia, China
PPSH-41 submachine gun (SMG)	7.62-mm submachine gun with 35- or 71-round magazines	Russia, China

Nomenclature[a]	Description	Origin(s)
Lanchester SMG	9-mm submachine gun, 25- or 50-round magazine	Great Britain
Thompson SMG	.45-caliber submachine gun, 20- or 30-round magazine	United States
Degtaryev light machine gun	7.62 drum-fed machine gun with 30-round magazine	Russia
PKM machine gun (type 80)	7.62-mm belt-fed machine gun	Russia, Warsaw Pact
Antipersonnel grenades		
Stick grenade	46-mm grenade with wooden handle	China
Antitank weapons		
RPG-2 antitank launcher (M-7)	40-mm handheld rocket launcher	Russia
RPG-2 antitank launcher (M-7)	40-mm handheld rocket launcher	Russia, Romania
Heavy weapons and antiaircraft weapons		
60-mm mortar	60-mm mortar with bipod	China, Vietnam
DShk antiaircraft machine gun	12.5-mm belt-fed heavy machine gun with 50-round canister	Russia
SA-7 Grail missile launcher	72-mm handheld antiaircraft missile	Russia
Armored vehicles		
T-34	Soviet medium tank, 76.2- or 85-mm gun, 26.5 tons	Russia
BTR-152	Soviet wheeled armored personnel carrier, 9.9 tons	Russia, Mozambique

[a] Chinese nomenclature in parentheses where applicable.

Eastern Bloc nations supplied both ZIPRA and ZANLA with armaments. These armaments were frequently surplus to Soviet and Warsaw

Pact stocks; heavy weapons and armored vehicles, when provided, were of World War II vintage. ZIPRA, as the Soviet client group, generally had the advantage of superior and more standardized weaponry over ZANLA, which often operated on a shoestring budget. Table 4-3 summarizes these weapons. As indicated in the table, the insurgents also used a hodgepodge of Western armaments obtained through commercial purchase.[160]

Tactical Re-evaluation

Despite the ultimate success of the insurgents in achieving majority rule, their initial military efforts in the mid-to-late 1960s were not fruitful, and this initial failure led to a reevaluation and change of tactics and strategy by both ZAPU and ZANU. This change in course would ultimately play an important role in leading to the electoral victory by ZANU and Robert Mugabe a decade later. Brickhill noted that ZAPU's military campaigns in the 1960s, which had mainly involved large, independent guerrilla detachments, had not been militarily successful, and by the early 1970s, it began to focus on ambush and land-mine warfare along the Zambezi river.[161] However, by the mid-1970s, ZAPU adopted a new military strategy designed to obtain a specific political goal—that of obtaining independence on its own terms. Specifically, key ZAPU officials argued that guerrilla warfare was a self-limiting strategy because it often just resulted in an insurgent movement gaining a seat at the negotiating table, rather than sweeping away a colonial power. Dumiso Dabengwa noted:[162]

> We were talking about seizing power. When we looked at other guerrilla wars we could see that guerrilla warfare does not enable you to seize power. It only creates the conditions for another force to settle the question. We wanted to take it a step further than that, and prepare ourselves to develop our military strategy and gear it to the final goal of a military victory. We felt that guerrilla warfare on its own could not achieve that. What is the next step?

The experience of the MPLA in Angola weighed heavily on ZAPU. The former had waged a guerrilla war that contributed to the collapse of the Portuguese government, yet the MPLA were almost defeated by the conventional military forces of South Africa.[163] It was only with the assistance of (conventional) Cuban forces that the MPLA was able to avoid defeat, a fact which was not lost upon Dabengwa:[164]

> The example of Angola was very fresh in our minds…
> We had realized that if we had to go through and take
> (our) country, we needed military forces that could
> seize power and defend it.

Into the late 1970s, ZAPU continued to rely upon raids, ambushes, and mine warfare, but it began to examine Vietnamese theories and practice of mobile warfare to determine how best to integrate conventional and guerrilla forces.[165] General Vo Nguyen Giap, the key architect of north Vietnamese military strategy against the United States, noted:[166]

> To keep itself in life and develop, guerrilla warfare
> has necessarily to develop into mobile warfare. This
> is a general law…If guerrilla warfare did not move to
> mobile warfare, not only the strategic task of annihi-
> lating the enemy manpower could not be carried out,
> but even guerrillas activities could not be maintained
> and extended.

The fundamental problem is that the absence of mobility led to the congregation of guerrilla forces within a particular geographic area. This in turn led to confusion among guerrilla forces and additionally made them easier targets for a conventional army. These problems were only compounded by the fact that the guerrillas typically lack the training and equipment to displace a conventional force.

These problems manifested themselves within the Rhodesian context. Dabengwa noted that by mid-1978 there was:[167]

> almost a sufficient presence of guerrilla forces in most
> ZIPRA operational areas. The danger in guerrilla
> warfare if you start having too many guerrilla units in
> an area, then you create confusion and lose the initia-
> tive…We did not think it wise just to pour in as many
> guerrilla units as possible without having specific
> objectives, just to have a presence in that area.

Additionally, a ZIPRA platoon commander noted:[168]

> By that time the unit which was operating on the northern side of the Khami river was in confusion. They were too many. They were conducting operations there, staying there. So the enemy located their positions and managed to see how they carried out their daily activities. Then they combed them.

> The introduction by the Rhodesian forces of airborne "Fireforce" tactics against the lightly armed guerrillas, was also seriously disrupting guerrilla efforts in the north of the country, where ZIPRA forces had grown most spectacularly. At this time the units at this front were also facing difficulties in mounting attacks against the heavily defended garrisons in to which most Rhodesian forces had retreated.

By the late 1970s, some ZAPU recruits had been diverted to conventional military training courses, and ZIPRA began to deploy regular forces armed with heavy equipment (such as various artillery pieces and anti-aircraft guns) with guerrilla units.[169, n] However, offensive and defensive deficiencies remained as guerrilla units lacked the infantry training to assault garrisons or to hold defensive positions.[171] It was within this context that Operation Zero Hour was conceived, which sought to effectively integrate conventional and guerrilla forces to sweep away the settler state. The plan called for a coordinated assault featuring five regular battalions with artillery support seizing northern bridgeheads at Chirundu, Kariba, and Kanyemba to permit ZIPRA armored reserves to cross and drive southward. Guerrilla units operating inside Rhodesia were assigned the mission to strike RSF and police installations, transportation and fuel storage facilities, and government offices on a broad front extending deep into the country. The ambitiousness of this plan is illustrated by ZIPRA's intention to seize airfields in Rhodesia so that its trained pilots could fly in MiG jet aircraft to support the offensive.[172] Moorcraft and McLaughlin claim that RSF cross border raids on ZIPRA delayed the planned execution of

---

n By 1979, ZIPRA stood up a conventional brigade featuring a variety of Soviet equipment, including T-34 tanks, MTU-55 bridging equipment, BTR-152 armored personnel carriers, recoilless rifles, field guns, heavy mortars, and command cars.[170]

Operation Zero Hour, although ultimately it was pre-empted by the Lancaster House talks.[173]

ZANU tactics and strategy also were subjected to significant scrutiny and subsequent changes. As previously noted (and discussed in greater length in the next chapter), various ZANU recruits were sent to China for training, where they were imparted lessons on the mobilization and politicization of the population. Beginning in the early 1970s, these lessons had a significant impact on the conduct of the war, particularly following the appointment of Josiah Tongogara, who was trained in China, as chief of ZANLA and ZANU secretary of defense.[174] Additionally, the arrival of Chinese instructors at the Itumbi camp in Tanzania in 1969 also played an important role in leading ZANU to adopt a Maoist "people's war" strategy against the Rhodesian state. Unlike ZAPU, ZANU never attempted a transition toward conventional tactics. Furthermore, most of its attacks occurred in rural areas.[175]

The kinetic aspect of this strategy played itself out in the following manner: once the inhabitants in the tribal trust lands were mobilized and politicized (which created the conditions for the insurgents to operate from a territory),[176] ZANU insurgents proceeded to attack any and all institutions and individuals (black or white) deemed to symbolize and represent the institutions of the minority government. Important targets included rural outposts of the Department of Internal Affairs (DIA) and its personnel, military, and police outposts and convoys, white farmers and their homesteads, tribal chiefs and headmen, and key infrastructure (e.g., railway lines and stations, telecommunication links, power lines, road, bridges).[177] Furthermore, these targets were attacked primarily through the use of landmines, surprise attacks, ambushes, and sabotage operations.[178]

Attacking the DIA was symbolically important for ZANU. This government department was in charge of administering the territory inhabited by the black African population, and the importance of attacking this symbol of governmental authority was noted by the political commissar of ZANLA:[179]

> All that had to do with Internal Affairs – those white District Commissioners and those black District Assistants – they were important targets. If we hit them it was in order to try to teach the people that the structure which supported the regime was vulnerable

to our attacks. The District Assistants were made to believe that the past regime was in no way going to be dismantled by anyone. They were also leading the people to believe that no black was ever going to be able to run district administration. You see, the DCs would go to the people and tell them that they are superior, they can crush the terrorists in no time. They used to gather the people at places like cattle dip tanks or at the Internal Affairs offices, and that's where they would tell the people that they had enough strength to attack terrorism. So it was these places that we attacked, to show people that what the DCs are saying is false.

Many of the DIA personnel killed by ZANLA were actually black Africans. Between 1974 and 1979, over 90 percent of the 304 killed were black.[180]

Attacks on the DIA were also intended to signal to the populace the credibility of ZANLA as an alternative source of authority. Similarly, attacks on military posts and convoys were intended to demonstrate the ineffectiveness of government security forces in rural areas.[181] Ambushes on convoys were often carried out by fifteen to twenty ZANLA guerrillas, with three to four armed with RPG-7s, and the rest armed with AK47s. Often the first vehicles were attacked with RPG-7s, or if the initial vehicles hit a landmine, the insurgents followed up with machine gun fire and RPG-7s.[182] Ambushes were also carried out against civilian travelers and commercial vehicles, and the intent was to demonstrate that ZANLA could attack at the time and place of its choosing and that the government was unable to protect travelers in rural areas.[183]

ZANLA also employed sabotage against a variety of targets, with the insurgents employing a modular method of deployment. Sabotage itself was typically carried out by a specially trained team of five to nine members who were supported by a larger unit consisting of fifteen to twenty combatants armed with machine guns, RPG-7s, and mortars.[184] The sabotage unit often used dynamite and grenades, and the supporting unit provided cover during an operation. Following the completion of a mission, the two units separated, and if the sabotage unit was assigned to a mission in another area, it would often be assisted by a different supporting unit.[185] Sabotage operations spread government forces thin and signaled to authorities that ZANLA could infiltrate tribal trust lands and use them to launch operations against the government.[186]

Popular Support for Resistance Movement or Insurgency

The Rhodesian conflict was characterized by segmented loyalties and tacit support of the black African population. Rooted in racial and ethnic conflict born out of colonial roots, the bonds of support for the Rhodesian government outside the white European minority were tenuous at best. "Coloreds"—mixed-race persons and Indian minorities—played little part in the outcome, although these groups were subjected to conscription by the government; black Africans serving in government positions, particularly in the RSF and the police, were all volunteers. ZAPU's base among Ndebele-speaking tribes and ZANU's base among Shona speakers is well understood. One Shona group, the Karanga tribe from Victoria Province, offers a poignant illustration of division among insurgents and government supporters, for it is from this tribe that the white Europeans had traditionally recruited native colonial troops and police. However, the insurgents also recruited successfully from this tribe, so it was not uncommon for a family to have relatives fighting on both sides.[187]

As noted previously, ZANU elicited popular support through *pungwes*, which were nocturnal rallies designed to politicize and mobilize the population. As one ZANLA political commissar noted:[188]

> [T]hese pungwes were intended to really politicize the masses. We used to take these lectures and teach them, explain to the masses stage by stage during the night...With all the regime's soldiers around, if it was done during the day that could mean disaster.

Upon arrival to a village, a combat unit would organize a *pungwe* with the help of *mujibas*, who were youths who assisted with the resistance effort. In particular, a newly arrived unit would consult with *mujibas* to ascertain the political attitudes of the local residents and to learn the movements of government forces. Additionally, they instructed the *mujibas* to inform local residents that a *pungwe* was to be held in the evening, with attendance mandatory. Those who did not attend were regarded as collaborators with the Smith regime.[189]

Pungwes served as a means to continually politicize and mobilize a village. They were held three or four times a week while a combat unit was in the area, and once the unit left, they recommenced once a new unit arrived within a week to ten days.[190, o] Security was provided by *mujibas* and guerrillas, who were typically stationed a kilometer away, and during the *pungwe* itself, ZANLA guerrillas led villagers in singing Chimurenga songs that emphasized various grievances and why the struggle was necessary. Participants often drank beer and ate food collected from local inhabitants.

Notably, participants were subjected to speeches from various junior-level commissars staffed with the combat unit organizing the *pungwe*. For instance, the assigned political commissar emphasized various grievances, and rather than discuss abstract political and ideological issues, he often emphasized tangible issues such as unequal access to fertile land. It was often noted that whites took their land and forced local residents to make do with substandard land in tribal trust lands.[192] Additionally, the political commissar noted that medical care was poor, and the education provided to village children was of low quality and would only prepare them for menial tasks.[193] This discussion typically reemphasized themes encoded in Chimurenga songs, and the political commissar would note that liberation would bring improved access to land, employment, and government services.[194]

ZAPU guerrillas frowned on *pungwes*, viewing them as "superficial and dangerous,"[195] and preferred to interact with the population through their clandestine party structures—a remnant from the early 1960s when ZAPU had been a legal political party. Both ZANLA and ZIPRA relied on mujibas.[p] Children as young as five years old could be *mujibas*. The ZANLA *mujibas* were motivated by the exploitation they observed of their parents and by the lack of adequate secondary

---

o Moorcraft and McLaughlin claimed that Zimbabwean guerrillas spent up to 80 percent of their time mobilizing the populace. While the authors did not indicate whether the guerrillas were from ZANLA or ZIPRA, it is clear that the authors were referring to the former because ZIPRA guerrillas did not attempt to politicize the populace. ZAPU interactions with the populace were handled by party officials rather than by the guerrillas.[191]

p Surprisingly, the term "mujibas" is of Slavic origin, as it is derived from the term "mujiva," which referred to individuals who were too young to join Marshall Tito's communist guerrillas in Yugoslavia during World War II. However, they still assisted in other ways, such as serving as messengers or interlocutors with the population and serving as the "eyes and ears" of the guerrillas.[196]

education in rural areas, and these themes were reinforced by political education they were subjected to by ZANLA commanders (ZANLA *mujibas* did not receive military training nor did they carry arms).[197]

Mujibas also acted as scouts, providing an early-warning mechanism for the guerrillas, and ran messages as needed. Regarding the former, *mujibas* sometimes used drums to provide advanced warning of enemy approaches, as one former ZANLA political commissar noted:[198]

> Well you know that saying: Action speaks louder than words. You'd find from the action of these mujibas – and their intelligence, of course – as to the way one reports to you on situations and developments. That would tell you if someone could be a good mujiba. Also, it was not only the way of doing reconnaissance that was important; it was also the way of giving the message. During the war it was important to give a message in a way in which one who is not a local of that area would not know what is taking place. That is why we sometimes used the drum. Say an enemy is approaching a village or an encampment where the guerrillas were nearby, then the drums could be a warning. There were different kinds of beats, so you could easily detect that, that beat means this thing. For the enemy's intelligence they relied on these walkie talkies and that was a problem to us. So often we discovered walkie talkies that had been given to the people so as to report on us. But our intelligence was better because we trusted the masses. In areas where the masses were united it would be very difficult for any enemy agents to infiltrate.

Mujibas also played a fundamental role in exchanging messages with combat units and, in the process, convey tactical and targeting information. Specifically, upon arriving to a village, a combat unit would contact *mujibas*, who would identify the combat unit based upon the coded message left behind by a previous unit, who would also leave behind instructions on targets and tactics along with information on arms caches and routes taken by ZANLA personnel.[199] *Mujibas* also provided ZANLA with information on the movement and strength of enemy forces and their direction of travel. Speaking of the *mujibas*, the defense correspondent for the Rhodesian Broadcasting Corporation noted:[200]

> Was their intelligence good? Amazingly so. Oh, shit –
> that mujiba system. It was those houts – mostly teenag-
> ers – who were ostensibly herding cattle, or whatever.
> Those houts, they knew exactly what was going on,
> exactly where the army was going and why. For exam-
> ple, they'd see a troopie loading beer crates – next
> thing you knew the word was back that the soldiers are
> off to drink beer. Yeah, they evolved a very, very good
> intelligence system with those mujibas.

The ZANLA *mujibas* also operated what was known as the "bush telegraph," by carrying messages over long distances, for instance from Salisbury to Umtali, a distance of 262 kilometers.[q] One official from the Rhodesian Intelligence Corp noted:[202]

> Their bush telegraph – that word of mouth network
> – was by far superior to our intelligence. They knew
> exactly what the Security Forces were doing, virtu-
> ally 24 hours a day, through their runners, the sym-
> pathizers, the mujibas. Those mujibas would give
> the *terrs* [terrorists] logistics, troop movements, troop
> strengths, and that was one of their greatest attributes
> as far as intelligence was concerned.

The female counterpart to a *mujiba* was known as a *chimbwidows*, and they often collected and cooked food for the guerrillas and washed their clothes. Yet above and beyond these more menial tasks, *chimbwid-ows* engaged in other gendered activities that assisted with collection efforts, such as serving as "honey traps" and consolers, as indicated by one ZANLA political commissar:[203]

> It was very difficult for us to know how the enemy forces
> were staying inside a camp or how they were perform-
> ing their duties; it was easier for the people to go and
> reconnoiter. For example, if it was a police camp, we

---

[q] *Mujibas* carried out other duties as well, such as carrying various war material for ZANLA guerrillas. They also collected food, soap, clothing, and blankets for the guerrillas and would advise ZANLA personnel not to accept food and clothing from certain indi-viduals, given concerns that donated items may have been poisoned.[201]

could send a woman – one of those women we called chimbwidows. We could send her, then she could be proposed by a policeman, and during their love affairs she can persuade him to tell her all the duties they carry out in the camp, how they are deployed, their ammunition, and eventually when that chimbwidow comes back to our base she can tell us all the information which she will have been given by that policeman. Then, automatically we will find it very easy to attack that camp. After that, when we have attacked the camp, some masses would go and try and feel pity for those soldiers or policeman who have been killed there and they would seem to be crying for their children. Yet they would be going to count the number of casualties. They can even help bury them but they would be counting the dead. Then they would come and tell us all the information they had learned.

They also carried war materiel, as noted by an official with the Rhodesian Intelligence Corps:[204]

And I'll tell you something – it wasn't just the young boys who were involved. They had a lot of women working with them. I remember this one time we were sitting on an OP one day and there was this junior troopie, sitting with me. I said: Do you notice anything unusual about the group of women – there were five of them – walking down the path? And he said: Yes, they're all carrying water down to the river. I said: Well, anything else? And he said no. I said: They're all pregnant, and they all look as though they're the same distance gone – don't you think there is something suspicious about that? So we went down to check it out. Land mines. Under their jumpers. It was incredible. And they're heavy, those things. Incredible. Five of the buggers. Ten land mines. One down their backs and one around the front, on rope. So I'd say that the women were pretty involved.

By 1979, as its reach across the Rhodesian countryside expanded, ZANLA claimed fifty thousand *mujibas*.[205] ZANLA was also able to foster mobilization and public acceptance of its activities through accessing and exploiting informal sources of legitimacy, such as spirit mediums and Shona tribal chiefs. A number of chiefs supported ZANLA, and such support facilitated various forms of popular assistance to the guerrillas. In particular, supportive chiefs instructed their followers to provide guerrillas with shelter, food, medicine, and intelligence, and once they obtained a chief's support, ZANLA insurgents were able to conduct themselves freely in areas controlled by the chief without fear of being reported to the security forces.[206]

However, it had been a long-standing practice of various Rhodesian governments to coopt the support of chiefs who, in return for salaries and insignias, dampened the fervor of black nationalism.[207] A noteworthy supporter of the government was Chief Jeremiah Sigireta Chirau, who along with Abel Muzorewa, Ndabaningi Sithole, and Ian Smith established the Executive Council in 1979. A decent number of chiefs supported a proposal in the early 1970s, which would have legitimized Rhodesia in exchange for a promise of eventual majority rule, a proposal which was largely rejected by most of the black population.[208] Yet of 184 (of 245 total) chiefs that were interviewed, forty-four supported the proposals (and eighty-seven rejected, with the remainder not answering or abstaining).[209]

One chief who opposed the government indicated that supportive chiefs provided valuable tactical information to the government:[210]

> The Smith regime only wanted chiefs of their choice so that they could become his puppets. Smith wanted chiefs who accepted money offers, since these people would not stand to oppose the government's policies. Instead they would connive with the government. The regime wanted chiefs who did not know their political rights, those who during the liberation war would locate the freedom fighters' bases so that they could be bombed. Certain chiefs did this because they were sell-outs. They were puppet chiefs. All those chiefs have been used by the regime at the expense of their own people. That is why I had to reject all the government's offers like a girl who rejects a boy's advances, at first sight and outright.

Spirit mediums were individuals through whom—it was believed—ancestors communicated to the living, and their importance in fostering legitimacy for ZANLA was noted by Bloch:[211]

> When the ZANLA guerrillas entered Zimbabwe, they realized that if they were to be successful they would have to be seen to be liberators by the people whom they had come to free. It was not in terms of the political analyses of socialist theoreticians that their actions would appear legitimate but in terms of the political ideas and interpretations of history of the peasants of Zimbabwe themselves...The Shona have always seen the relation between their past and their present as mediated by their ancestors. The young fighters therefore had to enter into a dialogue with these ancestors, to justify and explain their actions and to seek ancestral help.

Spirit mediums provided advice on where to travel and cache arms and even about strategy.[212] The most influential medium was that of Nehanda. The latter was herself a spirit medium who was hanged in 1898 and, before her death, indicated that her children would liberate the country.[213] The importance of Nehanda, and spirit mediums more generally, for mobilization was noted by Josiah Tungamirai, a ZANLA commander:[214]

> When we started the war the spirit mediums helped with recruitment. In the villages they are so powerful. If they tell their children they shouldn't go and join us they won't. When we wanted to go and open a new operational zone we would have to approach the mediums first. Mbuya Nehanda was the most important and influential recruit in those early days. Once the children, the boys and girls in the area, knew that Nehanda had joined the war, they came in large numbers.

Furthermore, Krieger noted:[215]

> The mediums and the guerrillas made a pact. The mediums would deliver peasant support and the guerrillas promised that if they were successful in war they

would reverse discriminatory legislation 'that limited the development and freedom of the peasantry' and, most importantly, they would return the land to the peasants. Thus began a process that culminated in the symbolic establishment of the guerrillas as the successors of the chiefs. 'Indeed they may be called the chiefs' legitimate successors because, like all legitimate rulers, they were installed by the *mhondoro* [spirits of deceased chiefs]. Apart from their recruiting of the peasantry to the resistance, the legitimization of this succession was the most important contribution the *mhondoro* mediums made to the war.' The mediums, by allowing 'this new feature in the experience of the peasantry,' the guerrillas, to be assimilated to established symbolic categories, facilitated the guerrillas acceptance by the population.

Derided as "witchdoctors" by Ian Smith, Pandya noted that Rhodesian authorities were too late in recognizing the importance of the spirit mediums:[216]

The white authorities, however, initially failed to understand the cultural beliefs of the black people and for the most part underestimated the significant influence the spirit mediums had on the population. As was the case with the tribal chiefs, whoever gained the support of the spirit mediums also secured the support of the population. ZANU and ZANLA, aware of their own cultural heritage, did not waste time in contacting and gaining the support of the spirit mediums. Once the spirit mediums had indicated their support for "the struggle," their followers automatically joined the campaign, thereby assisting ZANU and ZANLA in their mobilization process. As the campaign progressed, the authorities realized the importance of the spirit mediums and the following they had. So they "created" spirit mediums who would support them. These measures failed as it did not take long for the Shonas to identify the long-established spirit mediums from those "created" by the authorities.

In addition to the success of the *mujiba* network and its ability to gain the support of various spirit mediums and tribal chiefs, strong demographic trends assisted the insurgents' mobilization. Forty thousand black African Rhodesians came of age each year in the mid to late 1970s, and a substantial number of military-age youth had joined the ranks of displaced persons and refugees in the front line states. Others exfiltrated with assistance from insurgent networks to join the rapidly expanding ZIPRA and ZANLA forces in their training camps.[217]

The breadth of popular support for the insurgents was useful not only in helping them avoid detection by government security forces but also in identifying suspected government informers who might betray them.[218] Until 1972, the network of informers operating for the Rhodesian Special Branch had been extraordinarily effective in detecting and capturing ZANU and ZAPU guerrillas attempting to penetrate the cordon sanitaire along the border between Rhodesia and Zambia. The informers killed some insurgents and captured the rest so that they could be tried in court. ZANLA's incursion into northeast Rhodesia minimized the informers' effectiveness. The insurgents built a clandestine infrastructure for more than a year without being detected, using techniques learned from FRELIMO and ZAPU cadres who had reverted to ZANU in the wake of the Chikerema/FROLIZI schism of 1970. As Cilliers puts it, in 1972, the "Special Branch network of paid informers and police patrols . . . came close to total collapse in a matter of weeks."[219] ZANLA was gradually able to spread its influence southwest through the northeastern TTL toward Salisbury. By 1974, the Chiweshe TTL had become so dominated by ZANLA that the Rhodesian government moved its entire population of fifty thousand into protected villages.[220]

NOTES

[1] Carl Peter Watts, "The 'Wind of Change': British Decolonisation in Africa, 1957–65," *History Review* 71 (December 2011): 12–17.

[2] Ibid., 14.

[3] Ibid.

[4] Eliakim M. Sibanda, *The Zimbabwe African People's Union 1961–1987: A Political History of Insurgency in Southern Rhodesia* (Asmara, Eritrea: Africa World Press, Inc., 2005), 43–44, 58–61.

[5] Andre Astrow, *Zimbabwe: A Revolution That Lost Its Way?* (London: Zed Press, 1983).

6   David Martin and Phyllis Johnson, *The Struggle for Zimbabwe* (London: Faber and Faber, 1981), 51–54.

7   Ibid., 54–56.

8   Astrow, *Zimbabwe*, 8–10.

9   Martin and Johnson, *Struggle*, 56–59.

10 Alois Mlambo, "Building a White Man's Country: Aspects of White Immigration into Rhodesia up to World War II," *Zambezia* 25, no. 2 (1998): 125–126.

11 See Alois S. Mlambo, Some Are More White Than Others': Racial Chauvinism as a Factor in Rhodesian Immigration Policy, 1890–1963," *Zambezia* 27, no. 2 (2000): 143–144. See also Bizeck Jube Phiri, "The Capricorn Africa Society Revisited: The Impact of Liberalism in Zambia's Colonial History, 1949–1963," *International Journal of African Historical Studies* 24, no. 1 (1991): 65–83, for an elaboration of Rhodes's ideology in the context of CAF politics.

12 British Broadcasting Company, "Immigration and Emigration: Zimbabwe – or Was it Rhodesia," *Legacies*, accessed August 24, 2015, http://www.bbc.co.uk/legacies/immig_emig/england/gloucestershire/article_3.shtml.

13 Mlambo, " 'Some Are More White Than Others'," 142, 148, 150.

14 Sibanda, *Zimbabwe African People's Union*, 51–62, 88–90.

15 Timothy Scarnecchia, *The Urban Roots of Democracy and Political Violence in Zimbabwe* (Rochester: University of Rochester Press, 2008), 92–93.

16 Vladimir Shubin, *The Hot 'Cold War': The USSR in Southern Africa* (London: Pluto Press, 2008), 151–191. See also Christopher Andrew and Vasili Mitrokhin, *The World Was Going Our Way: The KGB and the Battle for the Third World* (New York: Basic Books, 2005), 439–444; and Scarnecchia, *Urban Roots*, 124–133.

17 Scarnecchia, *Urban Roots*, 112.

18 Ibid., 132–140.

19 Sibanda, *Zimbabwe African People's Union*, 96–97. See also Martin and Johnson, *Struggle*, 11; and Shubin, *The Hot 'Cold War*,' 151, 154–155.

20 Shubin, *The Hot 'Cold War*,' 151.

21 Ibid., 152.

22 Ibid.

23 Ibid., 154–155.

24 Anthony R. Wilkinson, *Insurgency in Rhodesia, 1957–1973: An Account and Assessment*, Adelphi Paper No. 100 (London: International Institute for Strategic Studies, 1973), 26.

25 Andrew and Mitrokhin, *World Was Going Our Way*, 444, 460.

26 David H. Shinn, "China's Involvement in Mozambique," *International Policy Digest*, August 2, 2012, http://www.internationalpolicydigest.org/2012/08/02/chinas-involvement-in-mozambique.

27 Martin and Johnson, *Struggle*, 23, 31, 71. Also see Shubin, *The Hot 'Cold War*,' 162.

28 Paul L. Moorcraft and Peter McLaughlin, *The Rhodesian War: A Military History* (Mechanicsburg, PA: Stackpole Books, 2008), 24, 130.

29 Astrow, *Zimbabwe*, 215–216.

30 Tom Meisenhelder, "The Decline of Socialism in Zimbabwe," *Social Justice* 21, no. 4, (1994): 83.

[31] Robert McKinnell, "Sanctions and the Rhodesian Economy," *Journal of Modern African Studies* 7, no. 4 (1969): 561–562.

[32] Martin and Johnson, *Struggle*, 14–20. See also Shubin, *The Hot 'Cold War,'* 171.

[33] J. K. Cilliers, *Counter-Insurgency in Rhodesia* (Dover, NH: Croom Helm Ltd., 1985), 119.

[34] Ibid., 174–195.

[35] Sibanda, *Zimbabwe African People's Union*, 161.

[36] Josiah Tungamirai, "Recruitment to ZANLA: Building up a War Machine," in *Soldiers in Zimbabwe's Liberation War*, ed. Ngwabi Bhebe and Terence Ranger (London: James Currey, 1995), 42.

[37] Jeremy Brickhill, "Daring to Storm the Heavens: The Military Strategy of ZAPU 1976 to 1979, in *Soldiers in Zimbabwe's Liberation War*, 52.

[38] Mao Tse-tung, "Problems of War and Strategy," in *Selected Works of Mao Tse-tung: Vol. II* (Peking, China: Foreign Languages Press, 1938), 224.

[39] Shubin, *The Hot 'Cold War,'* 180.

[40] Sibanda, *Zimbabwe African People's Union*, 24–25, 82–85. See also Martin and Johnson, *Struggle*, 44–50.

[41] Quoted in Martin and Johnson, *Struggle*, 27, 81.

[42] Sibanda, *Zimbabwe African People's Union*, 73.

[43] Shubin, *The Hot 'Cold War,'* 3.

[44] Andrew and Mitrokhin, *World Was Going Our Way*, 433, 440–442.

[45] Moorcraft and McLaughlin, *Rhodesian War*, 26–41, 62, 81–83, 103.

[46] Michael Kandiah and Sue Onslow, eds., *Britain and Rhodesia: The Road to Settlement* (London: Institute of Contemporary British History, 2008), 60–61.

[47] Baldwin Sjollema, *Never Bow to Racism: Personal Account of the Ecumenical Struggle* (Geneva: World Council of Churches Publications, 2015), 47.

[48] Moorcraft and McLaughlin, *Rhodesian War*, 82–83.

[49] Sibanda, *Zimbabwe African People's Union*, 103.

[50] Moorcraft and McLaughlin, *Rhodesian War*, 82.

[51] Martin and Johnson, *Struggle*, 10.

[52] Moorcraft and McLaughlin, *Rhodesian War*, 85.

[53] Yagil Henkin, "Stoning the Dogs: Guerilla Mobilization and Violence in Rhodesia," *Studies in Conflict & Terrorism* 36, no. 6 (2013): 505.

[54] Ibid., 507.

[55] Lebona Mosia, Charles Riddle, and Jim Zaffiro, "From Revolutionary to Regime Radio: Three Decades of Nationalist Broadcasting in Southern Africa," *African Media Review* 8, no. 1 (1994): 13.

[56] Compiled from multiple sources. Most of the biographical information on Zimbabwean African nationalist leaders has been extracted from an online version of Robert Cary and Diana Mitchell, *African Nationalist Leaders in Rhodesia – Who's Who*, accessed October 12, 2015, originally published in 1977. See http://www.colonialrelic.com.

[57] Sibanda, *Zimbabwe African People's Union*, 79–88; see also Martin Meredith, *The Past is Another Country: Rhodesia 1890-1979* (London: Andre Deutsch, Ltd., 1979), 316–317.

58 Scarnecchia, *Urban Roots*, 103–105. See also R. W. Johnson, "How Mugabe Came to Power: R.W. Johnson Talks to Wilfred Mhanda," *London Review of Books* 23, no. 4 (2011): 26–27; and Martin and Johnson, *Struggle*, 202–204.

59 Cary and Mitchell, "The Rev. Ndabaningi Sithole," *Who's Who*, accessed October 13, 2015, http://www.colonialrelic.com/the-rev-ndabaningi-sithole/.

60 Cary and Mitchell, "Jason Ziyapaya Moyo," *Who's Who*, accessed October 13, 2015, http://www.colonialrelic.com/106-2.

61 Cary and Mitchell, "Herbert Wiltshire Tfumaindini Chitepo," *Who's Who*, accessed October 13, 2015, http://www.colonialrelic.com/herbert-wiltshire-tfumaindini-chitepo. See also David Martin and Phyllis Johnson, *The Chitepo Assassination* (Greendale, Zimbabwe: Zimbabwe Publishing House, 1985), 38–59.

62 Cary and Mitchell, "James Robert Dambaza Chikerema," *Who's Who*, accessed October 13, 2015, http://www.colonialrelic.com/james-robert-dambaza-chikerema. See also Sibanda, *Zimbabwe African People's Union*, 72, 94; and Scarnecchia, *Urban Roots*, 105.

63 Reuters, "Lookout Masuku Dies at 46; Commanded Nkomo Forces," *New York Times*, April 7, 1986. See also Thulani Nkala, "Happy Birthday Lookout Masuku: Retracing His Footsteps," *Harare24 News*, April 6, 2012, http://harare24.com/index-id-Opinion-zk-13958.html; and " 'Zimbabwe, We Love You', As the Rebels Stream in from the Bush, Only Scattered Violence Mars the Truce," *Time*, January 14, 1980.

64 Cary and Mitchell, "Josiah Magama Tongogara," *Who's Who*, accessed October 14, 2015, http://www.colonialrelic.com/josiah-magama-tongogara. See also Mark Olden, "This Man Has Been Called Zimbabwe's Che Guevara. Did Mugabe Have Him Murdered?," *The New Statesman* (UK), April 12, 2004, http://www.newstatesman.com/node/195000.

65 Warren Foster, Nosimilo Ndlovu, Zodidi Mhlana, and Surika Van Schalkwyk, "The 'Black Russian' Changes Sides," *Mail & Guardian*, March 6, 2008, http://mg.co.za/article/2008-03-06-the-black-russian-changes-sides. See also "Dı Dumiso Dabengwa Profile - Zimbabwe's liberator.wmv," YouTube video, 12:43, posted by Zhou Media House, January 21, 2001, https://www.youtube.com/watch?v=UFHKgzI3gfA.

66 Andrew and Mitrokhin, *World Was Going Our Way*, 460.

67 Compiled from multiple sources. See *Tribute to Retired General Comrade Solomon Mujuru aka Comrade Rex Nhongo RIP* (blog), August 16, 2011, http://solomonmujuru.blogspot.com; and Moorcraft and McLaughlin, *Rhodesian War*, 37, 73–74, 114.

68 Lewis H. Gann and Thomas H. Henriksen, *The Struggle for Zimbabwe: Battle in the Bush* (New York: Praeger, 1981), 115.

69 Vladimir I. Lenin, "Report on the Unity Congress of the R.S.D.L.P.," in *Lenin Collected Works*, vol. 10 (Moscow: Progress Publishers, 1962), 376–381.

70 Sabelo J. Ndlovu-Gatsheni, "Nationalist-Military Alliance and the Fact of Democracy in Zimbabwe," *African Journal on Conflict Resolution* 6, no. 1 (2006): 56.

71 Ibid., 57. See also Moorcraft and McLaughlin, *Rhodesian War*, 85.

72 Moorcraft and McLaughlin, *Rhodesian War*, 85.

73 Ndlovu-Gatsheni, "Nationalist-Military Alliance," 56–59.

74 Paresh Pandya, *Mao Tse-Tung and Chimurenga: An Investigation into ZANU's Strategies* (Johannesburg: Skotaville Educational Division, 1988), 63.

75 Ibid., 64.

76 Ibid., 70.

77 Moorcraft and McLaughlin, *Rhodesian War*, 85.

78 Ibid., 86–87.

79 Pandya, Mao Tse-Tung and Chimurenga, 69–71.

80 Ibid., 70–71.

81 Ibid., 69.

82 Ibid., 71–72.

83 Dumiso Dabengwa, "ZIPRA in the Zimbabwe War of National Liberation," in *Soldiers in Zimbabwe's Liberation War*, ed. Bhebe and Ranger, 34.

84 Ibid., 26, 34.

85 Sibanda, *The Zimbabwe African People's Union 1961–87*, 166.

86 Peter Baxter, "ZAPU in the Zimbabwe Liberation Struggle," part 19 of 20 of the "History of the amaNdebele" series, *Peter Baxter: Author, Speaker & Heritage Guide* (blog), accessed October 14, 2015, http://peterbaxterafrica.com/index.php/2012/01/06/zapu-in-the-zimbabwe-liberation-struggle.

87 Moorcraft and McLaughlin, *Rhodesian War*, 70–71.

88 Cilliers, *Counter-Insurgency in Rhodesia*, 18, 28, 38, 51.

89 Martin and Johnson, *Struggle*, 308–309. See also Brickhill, "Daring to Storm the Heavens," 50–52.

90 Cilliers, *Counter-Insurgency in Rhodesia*, 51.

91 Martin and Johnson, *Struggle*, 276–278.

92 Cilliers, *Counter-Insurgency in Rhodesia*, 238.

93 Ibid., 241.

94 Scarnecchia, *Urban Roots*, 135.

95 Martin and Johnson, *Struggle*, 23. See also Amare Tekle, "A Tale of Three Cities: The OAU and the Dialectics of Decolonization in Africa," *Africa Today* 35, no. 3/4 (1988): 54–55.

96 Rhodesian Ministry of Foreign Affairs, *Communist Support and Assistance to Nationalist Political Groups in Rhodesia* (Rhodesia: Information Section, Ministry of Foreign Affairs, November 28, 1975), posted on *Rhodesia and South Africa: Military History* (blog), accessed August 22, 2014, http://www.rhodesia.nl/commsupp.htm.

97 Martin and Johnson, *Struggle*, 27. See also Dabengwa, "ZIPRA," 29–31.

98 "Zimbabwe's Liberation A Short and Accurate History," YouTube video, 47:20, posted by Antar Gholar, February 26, 2014, https://www.youtube.com/watch?v=z7iCjZf8ZGw.

99 Baldwin Sjollema, *Never Bow to Racism: A Personal Account of the Ecumenical Struggle* (Geneva: World Council of Churches, 2015), 97, 122.

100 Denis Herbstein, *White Lies: Canon Collins and the Secret War against Apartheid* (Oxford: James Currey, 2004), 262–271.

101 Ibid., 271.

102 H. Ellert, *The Rhodesian Front War: Counterinsurgency and Guerrilla War in Rhodesia 1962–1980* (Gweru, Zimbabwe: Mambo Press, 1989), 39.

103 Shubin, *The Hot 'Cold War,'* 151–154, 167–168, 180, 183.

104 Cilliers, *Counter-Insurgency in Rhodesia*, 19, 178, 180, 183–184.

105 Martin and Johnson, *Struggle*, 11–12, 24, 26–27, 83–84, 88. See also Sibanda, *Zimbabwe African People's Union*, 96.

[106] Sibanda, *Zimbabwe African People's Union*, 96. See also Moorcraft and McLaughlin, *Rhodesian War*, 74.

[107] Shubin, *The Hot 'Cold War,'* 171–172.

[108] Ibid., 73.

[109] Gann and Henriksen, *Struggle for Zimbabwe*, 50, 56–57.

[110] Sibanda, *Zimbabwe African People's Union*, 88.

[111] Andrew Novak, "Abuse of State Power: The Mandatory Death Penalty for Political Crimes in Southern Rhodesia, 1963–1970," *Fundamina* 19, no. 1 (2013): 35

[112] Moorcraft and McLaughlin, *Rhodesian War*, 81. See also Cilliers, *Counter-Insurgency in Rhodesia*, 40.

[113] Ibid. See also Baxter, "ZAPU in the Zimbabwe Liberation Struggle."

[114] Moorcraft and McLaughlin, *Rhodesian War*, 82.

[115] Mosia, Riddle, and Zaffiro, "From Revolutionary to Regime Radio," 12–13.

[116] Bruce Hoffman, Jennifer Taw, and David Arnold, *Lessons for Contemporary Counterinsurgencies: The Rhodesian Experience*, Report No. R-3998-A (Santa Monica, CA: RAND Corporation, 1991), 10.

[117] Martin and Johnson, *Struggle*, 276.

[118] Ibid., 267–277. See also Nathaniel Kinsey Powell, "The UNHCR and Zimbabwean Refugees in Mozambique, 1975–1980," *Refugee Survey Quarterly* 32, no. 4 (2013): 41–65.

[119] Tungamirai, "Recruitment to ZANLA," 40.

[120] Brickhill, "Daring to Storm the Heavens," 65.

[121] Ibid., 66.

[122] Ibid., 67.

[123] Ibid., 66.

[124] Ibid.

[125] Ibid., 67–68.

[126] Wazha G. Morapedi, "The Dilemmas of Liberation in Southern Africa: The Case of Zimbabwean Liberation Movements and Botswana, 1960-1979," *Journal of Southern African Studies* 38, no. 1 (2012): 76.

[127] Pandya, *Mao Tse-Tung and Chimurenga*, 77.

[128] Ibid., 85.

[129] Ibid., 77.

[130] Ibid., 79; and Tungamirai, "Recruitment to ZANLA," 40.

[131] Pandya, *Mao Tse-Tung and Chimurenga*, 79.

[132] Ibid., 79, 81.

[133] Ibid., 78.

[134] Gann and Henriksen, *Struggle for Zimbabwe*, 96–97.

[135] J. R. T. Wood, "Rhodesian Insurgency," Rhodesia & South Africa website, accessed November 10, 2014, http://www.rhodesia.nl/wood1.htm.

[136] Ibid., 90–93.

[137] Moorcraft and McLaughlin, *Rhodesian War*, 74.

[138] Sibanda, *Zimbabwe African People's Union*, 197–204. See also Brickhill, "Daring to Storm the Heavens," 61–65.

[139] Sibanda, *Zimbabwe African People's Union*, 203. See also Moorcraft and McLaughlin, *Rhodesian War*, 81.

[140] Wilkinson, *Insurgency in Rhodesia*, 31.

[141] Cilliers, *Counter-Insurgency in Rhodesia*, 37.

[142] Ibid., 6–9. See also Martin and Johnson, *Struggle*, 79–82.

[143] Moorcraft and McLaughlin, *Rhodesian War*, 100.

[144] Ibid., 100.

[145] Michael A. Stewart, *The Rhodesian African Rifles: The Growth and Adaptation of a Multicultural Regiment through the Rhodesian Bush War, 1965–1980*, Art of War Papers (Fort Leavenworth, KS: Combat Studies Institute Press, 2012), 34–38.

[146] Moorcraft and McLaughlin, *Rhodesian War*, 114–117. See also Brickhill, "Daring to Storm the Heavens," 62–63.

[147] Hoffman, Taw, and Arnold, *Lessons*, 10, 20.

[148] Baxter Tavuyanago, "The 'Crocodile Gang' Operation: A Critical Reflection on the Genesis of the Second Chimurenga in Zimbabwe," *Global Journal of Human Social Science* 13, no. 4 (2013): 27–36.

[149] Cilliers, *Counter-Insurgency in Rhodesia*, 7.

[150] Ellert, *Rhodesian Front War*, 7.

[151] "Rhodesia: Millions of Gallons of Fuel Destroyed as Guerrillas Blast Country's Largest Fuel Depot," December 12, 1978, http://www.itnsource.com/shotlist/RTV/1978/12/12/BGY511040193/, extracted October 26, 2015.

[152] Ibid.

[153] Sibanda, *Zimbabwe African People's Union*, 191–196.

[154] Moorcraft and McLaughlin, *Rhodesian War*, 154, 157.

[155] Tavuyanago, "'Crocodile Gang' Operation," 31.

[156] Ellert, *Rhodesian Front War*, 9.

[157] *The Murder of Missionaries in Rhodesia* (Rhodesia: Ministry of Information, 1978), http://www.archive.org/stream/TheMurderOfMissionariesInRhodesia/MMR_djvu.txt.

[158] Gann and Henriksen, *Struggle for Zimbabwe*, 98–99.

[159] Moorcraft and McLaughlin, *Rhodesian War*, 102.

[160] Ellert, *Rhodesian Front War*, 168–184. See also Moorcraft and McLaughlin, *Rhodesian War*, 92–98.

[161] Brickhill, "Daring to Storm the Heavens," *48–49*.

[162] Ibid., 49–50.

[163] Ibid., 50.

[164] Ibid., 52.

[165] Ibid., 50, 52.

[166] Ibid., 52.

[167] Ibid.

[168] Ibid., 53.

[169] Ibid., 50, 53.

[170] Moorcraft and McLaughlin, *Rhodesian War*, 96.

[171] Brickhill, "Daring to Storm the Heavens."

[172] Brickhill, "Daring to Storm the Heavens," 61.

[173] Moorcraft and McLaughlin, *Rhodesian War*, 77.

[174] Pandya, *Mao Tse-Tung and Chimurenga*, 175.

[175] Ibid., 173.

[176] Ibid., 176.

[177] Ibid., 163–165.

[178] Ibid., 163.

[179] Ibid., 167.

[180] Ibid., 168.

[181] Ibid., 168.

[182] Ibid., 171.

[183] Ibid., 170–171.

[184] Ibid., 171.

[185] Ibid.

[186] Ibid.

[187] Gann and Henriksen, *Struggle for Zimbabwe*, 71.

[188] Pandya, *Mao Tse-Tung and Chimurenga*, 145.

[189] Ibid.

[190] Ibid., 146.

[191] Moorcraft and McLaughlin, *Rhodesian War*, 100–102.

[192] Pandya, *Mao Tse-Tung and Chimurenga*, 147.

[193] bid.

[194] Ibid.

[195] Brickhill, "Daring to Storm the Heavens," 70.

[196] Pandya, *Mao Tse-Tung and Chimurenga*, 93.

[197] Ibid., 94.

[198] Ibid.

[199] Ibid., 95–96.

[200] Ibid., 96.

[201] Ibid., 95–97.

[202] Ibid., 97.

[203] Ibid., 99–100.

[204] Ibid., 100.

[205] Moorcraft and McLaughlin, *Rhodesian War*, 61, 72–73, 99, 104, 133–134.

[206] Pandya, *Mao Tse-Tung and Chimurenga*, 118–119.

[207] Ibid., 117.

[208] Luise White, "'Normal Political Activities': Rhodesia, the Pearce Commission, and the African National Council," *Journal of African History*, 52 (2011): 321–322.

[209] Pandya, *Mao Tse-Tung and Chimurenga*, 117.

[210] Ibid., 118.

[211] Maurice Bloch, "Preface" in David Lan, *Guns and Rain: Guerrillas and Spirit Mediums in Zimbabwe* (Berkeley, CA: University of California Press, 1985), xiii.

[212] Pandya, *Mao Tse-Tung and Chimurenga*, 127,129.

[213] Ibid., 129.

[214] David Lan, *Guns and Rain*, 147–148.

[215] Norma J, Krieger, *Zimbabwe's Guerrilla War: Peasant Voices* (Cambridge: Cambridge University Press, 1992), 129.

[216] Pandya, *Mao Tse-Tung and Chimurenga*, 130.

[217] Gann and Henriksen, *Struggle for Zimbabwe*, 98–99.

[218] Hoffman, Arnold, and Taw, *Lessons*, 10.

[219] Cilliers, *Counter-Insurgency in Rhodesia*, 221.

[220] Ibid., 73–74.

CHAPTER 5.
EXTERNAL SUPPORT

Although China and the Soviet Union were the main sponsors of the Zimbabwe African National Union (ZANU) and the Zimbabwe African People's Union (ZAPU), respectively, other countries played a supporting role in providing aid to the two liberation movements. A 1975 document released by the Rhodesian Ministry of Foreign Affairs noted that the training of small numbers of ZAPU and ZANU personnel also occurred in Bulgaria, Cuba, North Korea, Algeria, Ghana, Egypt, and Tanzania,[1] and the training of Zimbabwe People's Revolutionary Army (ZIPRA) personnel is alleged to also have occurred in East Germany[2,a] and Czechoslovakia.[4] Other East Bloc and non-aligned nations cited as having provided training and weapons to the insurgents include Yugoslavia, Romania, and Ethiopia.[5]

Tanzania, Zambia, Botswana, Mozambique, and Angola, the so-called "front line states," played vital roles in facilitating and enabling the insurgencies against the minority regime in Rhodesia. The two insurgent groups maintained offices and training camps in neighboring countries and launched operations from these territories, and vital supply routes passed through neighboring countries. In the case of ZANLA, the organization maintained three types of camps in neighboring countries: base camps, transit camps, and staging posts.[6] A base camp housed approximately twenty thousand to thirty-five thousand individuals and was run by a camp commander, who was assisted by a political commissar and education, logistics, and medical officers.[7] The political commissar was responsible for the political training of the camp population and, in particular, making sure the latter was "woke" with respect to the oppression and inequality experienced by the black population in Rhodesia. The entire camp population was divided into groups of fifty or sixty people for political training, which occurred every day for two hours.[b]

Transit camps were much smaller and served a dual purpose. First, guerrillas heading into Rhodesia were issued supplies at such camps,

---

[a]   The document released by the Rhodesian Ministry of Foreign Affairs also noted that East Germany printed and circulated the ZAPU newsletter, *The Zimbabwe Review.*[3]

[b]   Additionally, the security officer for the camp made sure that the camp's water supply was not poisoned, collaborators had not infiltrated the camp, and sentries were posted a half kilometer away from the camp. Furthermore, the education officer was responsible for youth education in the camps, with syllabi developed by ZANU's education committee, and textbooks were published by the printing division at ZANU's Publicity and Information Department in Maputo, Mozambique.[8]

and they served as a rest stop for guerrillas returning from Rhodesia.[c] Staging posts represented the last point of debarkation for ZANLA personnel heading into Rhodesia, and at these locations, guerrillas obtained instructions on their deployment and actions to take once inside Rhodesia.[10]

Of note is that at varying times different regional leaders could effectively threaten to shut down the liberation struggle or otherwise significantly impact the campaign against Rhodesia if they disagreed with ZANU or ZAPU leaders' decisions on strategy or organizational matters or if the conflict was deemed detrimental to national interests. For instance, factionalism (both within and between the two insurgent groups) was a persistent complaint of regional leaders. In the early 1970s, Kenneth Kaunda threatened to expel ZAPU from Zambia after infighting broke out at ZAPU camps in Zambia.[11] Additionally, in the mid-1970s, after the imprisoned ZANU Central Committee dismissed Ndabaningi Sithole from his leadership position for denouncing violence, Samora Machel, the head of Frente de Libertação de Moçambique (Mozambique Liberation Front, or FRELIMO) and the first president of Mozambique, threatened to shut down the war effort if Sithole was deposed.[12,d]

Another important event that demonstrated the insurgents operated in neighboring territory only on the sufferance of regional leaders was the 1975 assassination of Herbert Chitepo in Zambia. Suspicion immediately fell on ZANU. Zambia imprisoned nearly every member of ZANU's leadership and the high command of the Zimbabwe African National Liberation Army (ZANLA) after the assassination, and ZANU's offices in Zambia and Tanzania were shuttered. Additionally, ZANU military recruits in Mozambique were disarmed and isolated from Robert Mugabe and Edgar Tekere, who were ZANU's only political leaders in Mozambique.[13] Lastly, after the devastation wrought on regional countries by the Rhodesian armed forces in the late 1970s, regional leaders put significant pressure on the leaders of ZAPU and ZANU to reach an agreement at the Lancaster House talks in 1979.[e]

---

[c] A section typically served six to eight months within Rhodesia before returning to a transit camp for rest and to replenish supplies.[9]

[d] At this time, many insurgent operations against Rhodesia were launched from Mozambique.

[e] There are indications that regional leaders issued an ultimatum to the leaders of ZANU and ZAPU suggesting that they would not be able to return if they did not reach an

Additionally, both the Soviet Union and the People's Republic of China (PRC) often had to seek the approval of regional actors before lending assistance to ZAPU and ZANU. For instance, China had to obtain permission from President Julius Nyerere of Tanzania to permit Chinese personnel to train insurgents in Tanzania,[15] and after host countries exhibited opposition at a 1975 military conference in Mozambique, the Soviet Union began to minimize its role in support of Zimbabwe People's Army (ZIPA).[16] Hence, given the important role played by regional countries and other actors, such as the Organisation of African Unity (OAU), before discussing Soviet and Chinese sponsorship of ZAPU and ZANU, respectively, the next section discusses the role played by various regional actors.

THE ORGANISATION OF AFRICAN UNITY

The OAU formed in 1963, and its overarching role was to present a Pan-Africanist[f] face to the world and lobby for international support in the ongoing struggles for independence and majority rule in countries across the continent. In Africa, it became the legitimizer of nationalist movements, and in an effort to promote unity among national liberation movements within a conflict, it generally recognized a single nationalist movement. In the case of Rhodesia, that entity was ZAPU. A critical component of the OAU was its Liberation Committee, based in Tanzania. The Liberation Committee, which worked under the guidance of a governing board of OAU member states, closely collaborated with the government of Tanzania and its designated officials and structures to provide funding, logistical support, training, and publicity to all recognized liberation movements. Kempton noted that although OAU funding was typically modest, more important was the legitimacy and access to the territories of member states that official OAU

---

agreement in London.[14]

   [f] Pan-Africanism has been described as "an intellectual and political outlook among African and Afro-Americans who regarded Africans and people of African descent as homogeneous. . . . This outlook led to a feeling of racial solidarity and a new self-awareness and caused Afro-Americans to look upon Africa as their real 'homeland,' without necessarily thinking of a physical return to Africa." As a political movement, it promoted the political unity of Africa and emphasized the cultural unity of the continent. Its origins can be dated to the first Pan-African congress in London in 1900, which was attended by approximately thirty delegates, primarily from England and the West Indies but also including a few African Americans.[17]

recognition conferred. Additionally, the leaders of the African Front Line States, particularly Tanzanian President Julius Nyerere, sought to exercise leverage over the Zimbabwean insurgents by attempting to control the flow of weapons and training through the OAU Liberation Committee.[18]

TANZANIA

President Julius Nyerere was a vigorous supporter of majority rule in Rhodesia. At the start of December 1974, talks in Lusaka involving nationalist leaders and Rhodesian and South African officials, he stated:[19]

> We are not here to discuss whether there should be majority rule before independence. Neither should we discuss majority rule in ten or five years, but immediate majority rule. All that is negotiable should be about how, by what steps, and with what timing independence on the basis of majority rule will be established. If the principle of majority rule is not accepted then the basis for a Rhodesian constitutional conference does not exist. There will be no cessation of the guerrilla war until Smith crosses the Rubicon by accepting this principle.

Nyerere backed up such statements by making Tanzanian territory available to train insurgents. ZANLA operated a camp at Itumbi, which opened in 1965 and had the capacity to train approximately two hundred insurgents, and in 1969, it featured Chinese instructors.[20] In 1971, Itumbi was replaced by a camp at Mgagao, and during the latter parts of the war, Mgagao served to train instructors who were then deployed to camps in Mozambique to train combatants.[21] Meanwhile, ZAPU operated a camp at Morogoro that included Soviet instructors.[22]

Tanzanian officials also worked closely with their Zambian counterparts to establish other camps in Tanzania as training sites for insurgents from Zimbabwe. In early 1967, it was decided to utilize a camp at Kingolwira to train Zimbabwean recruits on a rotational basis, with one hundred ZAPU insurgents undergoing training followed up by a similar number of ZANU recruits.[23] Another camp was set up at Kongwa while a transit camp was set up in Chunya for insurgents returning from training and awaiting deployment.[24]

Officials from the two countries also arranged for the transport of Zimbabwean recruits from Francistown in Botswana to camps in Tanzania via Zambia. These recruits entered Zambia by first traveling to Botswana because the Rhodesia–Zambia border became highly militarized following the Unilateral Declaration of Independence (UDI). Once the insurgents were inside Zambia, Zambian authorities informed their Tanzanian counterparts to indicate the number of militants ready for training in Tanzania. Once the required recruitment forms were filled out, the militants entered Tanzania through the Nakonde–Tunduma border point, and they returned to Zambia through the same location.[25]

ZAMBIA

President Kenneth Kaunda took an obvious interest in the Rhodesian conflict given Zambia's long border and history with Rhodesia. Immediately following independence, Kaunda indicated that Zambia would support the struggle for majority rule in southern Africa:[26]

> The basic aim of Zambia's foreign policy is to secure peace, freedom and prosperity through justice at home and to maximize our contribution to world peace and the welfare of mankind . . . Under our policy . . . we cannot hold our heads high before the rest of the world unless we take our full part in helping those of our brothers and sisters [in southern Africa] currently struggling to free themselves from racial oppression and minority exploitation. We shall continue to give them all the support we can.

This stance though did not simply reflect one of principles. It also reflected the view among Zambia's leaders that the country's continued independence was inimical with the continuance of colonial and minority rule based upon racial hierarchy in surrounding countries.[27] However, as was the case with Botswana (see below), Zambia had to tread carefully because its economy was highly integrated with Rhodesia's. For instance, in 1965, nearly its entire bilateral trade through the ports of Beira (in Mozambique), Lourenço Marques (present-day

Maputo in Mozambique), and various South African ports was transported by Rhodesia railways.[28] In fact, its southern neighbor purchased 93 percent of its exports and supplied 33 percent of Zambia's merchandise imports.[29] Additionally, Rhodesia supplied much of the country's energy needs, including 95 percent of its coal requirements, 90 percent of its oil and petroleum products, and much of its electric power (from the Kariba South Bank hydroelectric power station).[30]

However, the country's leadership concluded that Zambia's economy would always remain vulnerable as long as Rhodesia, Angola, and Mozambique were ruled by colonial powers or minority regimes, and hence support for national liberation movements in those countries was seen as consistent with the country's long-term economic interests.[31] Additionally, Zambia supported various liberation movements to not lose face with other African countries, as indicated in 1965 by one cabinet official:[32]

> Withdrawing our support from the freedom fighters would be in conflict with the avowed aims of the OAU and other self-respecting states in Africa, apart from being in violation of our fundamental principles upon which Zambia was founded. Such actions would also make Zambia one of the most sinister nations that have ever polluted the pages of history of the independence movements in Africa. Zambia has the moral duty to help in the historic movement of wiping out colonialism in favor of the democratizing process.

The country provided various forms of support to both ZAPU and ZANU. In 1965, the government purchased a new office building in Lusaka to house the African Liberation Center, which provided office space to a variety of liberation movements. The director of the center worked closely with the OAU Liberation Committee to funnel recruits to training camps.[33] The country also provided right of passage and transportation facilities for the transport of insurgents to training camps in Tanzania and outside of Africa, as well as the shipment of war materiel.[34] It also served as a transit point for insurgents returning from training in other countries,[35] and the government made available broadcasting equipment for use by Zimbabwean insurgents. In 1966, ZAPU was permitted to broadcast from the Zambia Broadcasting

Services, while in the same year, ZANU was permitted to use the facilities at Radio Zambia.[36]

Furthermore, while there were a number of insurgent camps in the country, in general Zambian authorities did not permit insurgent training in the country owing to fears of Rhodesian reprisal attacks.[37] Instead, as discussed in the previous section, Zambian officials worked closely with their Tanzanian counterparts to establish training camps in Tanzania. There were, however, a number of transit camps in Zambia, prominent among which were those at Chikumbi, Mboroma, Mkushi, and Lusaka West.[38] Additionally, Pandya noted that the Chifombo camp on the border with Mozambique was a ZANU camp that housed between nine thousand and thirteen thousand people, and it served as a transit camp and training site.[39]

The latter fact indicates that some insurgent training did indeed occur in Zambia. Chongo noted that some training also occurred at Chikumbi.[40] In any case, Rhodesia launched a number of air attacks against various camps in Zambia. In October 1978, Rhodesia attacked Chikumbi and Mkushi, with hundreds killed in the operation. Mkushi served as a training site for girls, and there are indications that ZAPU's main communications center was housed at Chikumbi and that various weapons (including SAM-3 anti-aircraft missiles and surface-to-surface rocket launchers) were stored at the camps.[41] Other attacks included a March 1978 raid on a suspected ZIPRA camp in Kavalamanja village in Luangwa District; a November 1978 aerial bombing of a suspected ZAPU base west of Lusaka; aerial bombings of ZAPU refugee camps in February and April 1979; and the bombing by Rhodesian warplanes of a ZAPU guerrilla camp in Kalomo.[42]

These and other attacks were intended to prevent infiltration into Rhodesia from Zambia. Interestingly, infiltration across the Zambezi river had early proved problematic for ZANLA, as various factors inhibited the use of the Rhodesian territory on the other side of the river as a rear area for further infiltration deeper into Rhodesia. Tongogara noted that:[43]

> They [the insurgents] found few people in the area
> on the Rhodesian side who could provide them with
> food, shelter and information. The combatants walked
> for between two and five days through forbidding and
> thick bush before they reached the first villages. In this

area there was a shortage of water, it was excessively hot, dangerous animals were many and most important, the security forces, who saw the main threat as coming from Zambia, had set up a cordon sanitaire with camps along the Rhodesian side of the river.

Such difficulties led Tongogara and other ZANLA commanders to conclude it was necessary to set up another front in the Tete province in Mozambique:[44]

> We realized that we could not carry on because when you are in a revolutionary war you first have to have a strong rear base and then you endeavor to create a strong front. These things have to work cohesively. If the rear is weak the front cannot be strong. So we discovered that we were doing a very good job inside, according to that time – it was the beginning – but the rear was weaker than the front, despite the hardship of the front. We discovered that we could not continue with the war like that and win. So ZANU and ZANLA decided they must go through Tete.

Rhodesia took other punitive measures against Zambia. In January 1973, Rhodesia closed its border with Zambia to punish it for allowing its territory to be used to launch attacks against Rhodesia. This move forced Zambia to seek alternative routes and sources for the nine hundred thousand tons of imports and four hundred thousand tons of exports that passed through the Rhodesia–Zambia border.[45]

Lastly, it is important to note that Kaunda was a leading force behind seeking a unification of the ZAPU–ZANU nationalist effort. He strongly felt that tension between ZAPU and ZANU was counterproductive and the energy spent quarreling with each other could be much better spent fighting the Smith government. Partly because of ZAPU's preponderance of official recognition across Africa and the policy of détente toward South Africa in 1976, Kaunda favored ZAPU and its leader, Joshua Nkomo.

ANGOLA

Following its independence from Portugal in the mid-1970s, Angola began to play a prominent role in the conflict in Rhodesia, particularly with respect to ZAPU. In 1977, Angolan authorities agreed to the establishment of a ZIPRA training camp near Luena, in eastern Angola.[46] The camp was staffed by both Soviet and Cuban military personnel, and the former eventually consisted of specialists in tactical training, small arms, engineering, topography, and fire-range equipment. The Soviet contingent was at one point led by Lt. Col. Vyacheslav Zverev, who had been a commander of a training battalion in the Turkestan military district, and the training program was developed in coordination with the Cubans.[47] Ben Matiwaza was the ZIPRA camp commander, and every two months, two thousand ZAPU personnel would arrive from Zambia.[48] In late February 1979, Rhodesian and South African aircraft bombed the camp, resulting in the deaths of 192 ZAPU fighters, and hundreds more were injured. Also among the dead were six Cuban instructors and one Soviet warrant officer.

Angola also served as a transshipment point for Soviet aid to ZIPRA. In particular, Mavhunga noted that beginning in July 1978, convoys of armored vehicles left southern Angola for Zambia. Additionally, ZIPRA personnel trained in Angola had use of T-34 tanks, MTU-55 bridging equipment, BTR-152 armored personnel carriers, and BM-14 and BM-21 multiple-rocket launchers.[49]

BOTSWANA

Botswana gained its independence in 1966, and from the start, it was placed in a difficult position with respect to the conflict in Rhodesia. It was essentially surrounded by white minority regimes in South Africa, Rhodesia, and southwest Africa (which became Namibia in 1990), and it was economically dependent on the first two countries. In particular, it relied on South Africa and Rhodesia for the importation of manufactured goods, cereal grains, and other important foodstuffs, while the two countries also were important markets for beef and other animal products from Botswana.[50] Notably, Rhodesian railways ran through Botswana but was owned and managed by Rhodesia, and Botswana was reliant on the railway to transport its goods to seaports in South Africa. This dependence led a one-time foreign minister of Botswana to claim that the country's regional geopolitical context was comparable to being "under the belly of a whale."[51]

Hence, shortly after independence, Botswana took a realistic approach toward the brewing insurgency in Rhodesia, as the country did not allow its territory to be used as a refuge or base of operations for Zimbabwean resistance movements. This stance was communicated by President Seretse Khama in a speech to the National Assembly six days after independence:[52]

> My government will not interfere in the internal affairs
> of other countries and will not tolerate interference in
> Botswana's affairs by other countries. In particular, we
> will not permit Botswana to be used as a base for the
> organization or direction of violent activities directed
> against other states and we will expect reciprocal treat-
> ment from our neighbors.

Botswanan fears were crystalized following the August 1967 engagement involving thirty-three ZAPU and South African African National Congress (ANC) insurgents and Rhodesian troops, which saw the insurgents use Botswana as a refuge. Both South Africa and Rhodesia threatened reprisals and a hot pursuit of the guerrillas, which prompted Botswana to round up and deport the insurgents to Zambia.[53] However, by the mid- to late 1970s, thousands of Zimbabweans, many of whom were ZAPU recruits, streamed into Botswana, with inflows increasing from 5,712 in 1976 to 25,300 in 1978.[54] These entrants were accommodated at transit centers in Selibe Phikwe and Francistown and at a refugee camp at Dukwi.

The discovery of diamonds in the country in 1969 and the commencement of production two years later enabled the country to develop a degree of economic independence from South Africa, which was evidenced by its signing of the April 1969 Lusaka Manifesto, which acknowledged the use of violence as a legitimate final option for black majorities to gain independence in southern Africa.[55] Nonetheless, the country did not permit the presence of ZAPU or ZANU training camps on its territory, although it allowed insurgent movements to set up administrative offices in the country. By the late 1970s, the government turned a blind eye toward the activity of insurgent groups in the country, as it grew tired of persistent border skirmishes and Rhodesian forays into the country.[56]

SOUTH AFRICA

In the early to mid-1960s, South African strategists were primarily concerned with the "black peril" of African nationalism, rather than the "red menace" of communism.[57] Surrounding territories were seen as a buffer against black-ruled states. This threat perception changed with the appointment of P. W. Botha as minister of defense in 1966. The new minister located the threats to South Africa within a cold war context and, in particular, viewed South Africa as playing an important role in countering the Soviet threat to the West.[58] Hence, a buffer was still needed, but now to keep communist forces and influences as far away as possible from South Africa.

South Africa began providing military assistance to Rhodesia in 1967, when South African paramilitary police were deployed into Rhodesia to assist with combat against ZAPU forces and the South African ANC in the Wankie and Sepolilo campaigns.[59] Additionally, between January 1968 and August 1975, South Africa provided a variety of equipment to Rhodesia, including various aircrafts (Vampires, Dakotas, and Alouette II helicopters), weapons (rifles, shotguns, machine guns, rocket launchers, cannons, mortars, etc.) and vehicles.[60] In the mid-1970s, South Africa was also deploying various personnel, including helicopter pilots and technicians, and was providing assistance in the construction of five new military airfields. By 1978, South Africa was also deploying troops in southern Rhodesia, and in 1979, it was supporting Rhodesian raids into Zambia and Mozambique.[61]

South Africa also provided substantial financial and economic assistance to Rhodesia. In 1975–76, it financed 50 percent of the Rhodesian defense budget, and in 1979, the *New York Times* reported that South Africa was covering 40 percent of the $2 million per day cost of the conflict.[62] South Africa also did not enforce the comprehensive economic sanctions imposed by the United Nations on Rhodesia.[63]

MOZAMBIQUE

In conjunction with South Africa, the Portuguese colonial government in Mozambique acted in defiance of the international community by ignoring the sanctions regime. This allowed the Smith government to continue to have access to its critical oil supplies and to the port

151

facilities in Maputo for shipping its exports to willing buyers around the world. The colonial government in Mozambique waged its own campaign against FRELIMO, near the border with Rhodesia, and its alliance with Rhodesia provided a fair degree of security cooperation between the two countries along the eastern border.

In 1974 a military coup in Portugal caused a collapse of colonial governments in Mozambique and Angola, which drastically changed the economic and security situation facing the Rhodesian government. FRELIMO came to power in Mozambique. While the Rhodesian government would no longer enjoy the benefits of an ally along their eight-hundred-mile shared border, it also had to contend with a Soviet proxy that was willing to support Rhodesian liberation movements.

In particular, the fall of the colonial government in Mozambique paved the way for ZANLA to establish a variety of camps in the country and to launch operations from various Mozambican provinces. ZANLA guerrillas began to operate from the Tete province in 1972, and the fall of the colonial government enabled ZANLA to operate from staging areas in Mozambique along the entire border with Rhodesia. Furthermore, after 1974, most ZANLA recruits were trained in Mozambique,[g] and notable base camps included the Mavudzi, Doroi, and Tronga camps, while other base camps included Chimoio, Tembwe, and Nyadzonya.[65] Transit camps included the Chicualacuala, Chikombidze, Madulo Pan, Matimbe, Rio, and Mahiba camps, and Caponda and Mazaila served as staging posts.[66] ZANLA launched operations from three Mozambican provinces, Tete, Gaza, and Manica, and these operations were further decomposed into operational sectors that overlapped with tribal trust lands throughout Rhodesia.

Training in Mozambique lasted six months, although in Tanzania, it lasted sixteen months.[67] Recruits in Mozambican camps were largely instructed by trainers trained in China (and Tanzania) and were subjected to substantial political training during the first month, as they were made aware of the key grievances motivating the conflict. Additionally, they were instructed in the works of Marx, Engels, and Mao, and ZANU established the Chitepo College of Political Ideology at Chimoio to train political commissars.[68] Regarding military training,

---

[g] Toward the end of the war, there were thirty thousand to thirty-five thousand ZANLA personnel in Mozambique, as well as one-hundred-fifty thousand refugees from Rhodesia.[64]

recruits underwent instruction on the use of weapons useful for a protracted guerrilla war, including automatic machine guns, rifles, mortars, bazookas, pistols, and light machine guns, as well as the placement of land mines and anti-personnel mines. As Pandya noted, "They [ZANLA] had neither the equipment (such as tanks, artillery, helicopters, fighter aircrafts, air to air, and ground to air missiles) nor the trained personnel to develop the protracted war into a conventional war."[69]

The tribal trust lands along the border with Mozambique proved to be fertile grounds for subsequent infiltration deeper into Rhodesia, as the local population proved to be sympathetic to ZANU's campaign against the minority government. In particular, the local populace provided ZANU insurgents with information, food, and shelter.[70, h] Additionally, the geographic terrain facilitated insurgent activity. The north and northeastern regions of Rhodesia featured numerous rivers, along with the Rukowaluona and Mavuradonha mountain ranges, and caves in the mountains offered hiding places and locations to cache arms. Additionally, the numerous rivers supported thick green forests to facilitate undetected movement.[72] Similar geographic features in the Eastern region of Rhodesia facilitated infiltration from Mozambique.[73]

THE SOVIET UNION

Actors and Methods Used to Provide Support

A variety of Soviet actors had a hand in providing support to ZAPU, as evidenced by Joshua Nkomo's itinerary during a January 1976 visit to Moscow, which included meetings with the Soviet Afro-Asian Solidarity Committee; a meeting with Rostislav Ulyanovsky, the First Deputy Head of the International Department of the Communist Party of the Soviet Union (CPSU) Central Committee; a meeting at the Ministry of Defense with the *Desyatka*, or the Tenth Main Department of the General Staff of the Armed Forces of the Soviet Union, which managed Soviet military cooperation with foreign countries; and a meeting at the Africa Institute in Moscow, an organization established by the noted

---

h Conversely, independent Mozambique served as an ideal sanctuary for ZANU and ZANLA, given the common language and culture Mozambique shared with Shonas in Rhodesia.[71]

Soviet Africanist, Ivan Potekhin, who also at one point was the chairman of the Soviet Association of Friendship with African countries.[74]

Perhaps the most important actor was a party organization rather than a state body, the International Department (ID), which was founded in 1943 around the time Comintern was disbanded.[75] Under the leadership of Boris Ponomarev, who led the department from 1955 to 1986, the ID took an active role in working with radical movements in the Third World as well as sponsoring front organizations.[76] The pivotal role played by the ID is nicely captured by Shubin:

> The process of decision-making in Moscow regarding the Zimbabwe liberation movement was as follows. Soviet embassies in several African countries, primarily Tanzania, then Zambia and finally Angola, were in regular contact with ZAPU leaders and representatives. They had diplomats among their staff responsible for contacts with 'NOD' (a Russian abbreviation for 'National Liberation Movements'). Most 'sensitive' correspondence, including the requests for supplies, went via diplomatic post, but in urgent cases a letter would be preceded by 'shifrovka', a coded telegram. Otherwise those requests would often be brought to Moscow by liberation movements' own delegations. Either way, requests would initially be looked through in the CPSU international department, the staff members of which would draft the decision of the Central Committee, which instructed state bodies to consider the request. And finally the proposal drafted by those bodies (in liaison with the international department) would be considered by the highest body, the Politburo.[77]

For instance, in mid-1976, Nkomo forwarded a written request for Soviet arms to Professor Vassily Solodovnikov, the Soviet ambassador to Zambia and prominent Soviet Africanist who previously headed the Africa Institute.[78] Ambassador Solodovnikov passed the request to the CPSU, which in turn directed the Ministry of Defense and other state organizations to submit their plans for fulfilling the request.[79]

This episode highlights the bifurcation of the Soviet foreign policy bureaucracy into party and governmental organizations, and at times the different missions of such bodies led to interorganizational tensions. Whereas the ID had a mandate to support revolutionary movements, the Soviet Ministry of Foreign Affairs often wanted to improve ties with existing governments,[80] and in addition the two organizations sometimes disagreed over which liberation movement to support in Zimbabwe. Petr Yevsukov, the Soviet ambassador to Mozambique, noted:[81]

> From Maputo the balance of forces in Rhodesia . . . was seen better. It was clear that Mugabe enjoyed the support of the majority of the African population as well as of Mozambique, especially President Samora Machel. . . . The necessity of amending our policy with regard to support of the forces of national liberation in Zimbabwe, taking into account the likelihood of Mugabe becoming the leader of the independent country, became evident for the Soviet embassy in Mozambique. . . . To begin with, I invited Mugabe for a meeting. The discussion took place in my residence, which was almost next to the mansion where Mugabe lived. I communicated the content of our talk and the embassy's proposal to Moscow. . . . Soon after the meeting with Mugabe I flew to Moscow, when I received support on this issue from a number of influential people and institutions, including L. F. Ilyichev, Deputy Minister of Foreign Affairs, and Petr Ivanovich Ivashutin, Chief of the GRU of the General Staff. But at the discussion in the CC International Department, I faced resolute opposition from R. A. Ulyanovsky. . . . Having got angry, he said: "Why have you met Mugabe? Nobody instructed you to do so." Such a position resulted in me feeling awkward, when Zimbabwe became independent and Mugabe became its first Prime Minister.

Efforts by Dr. Venyamin Chrikin, a Soviet legal advisor sent by the ID to provide advice to ZAPU, to establish ties to Mugabe were also rebuffed:[82]

> According to instructions received . . . in Geneva and later in 1979 in London we could only have contact with Nkomo's party. In our messages to the CPSU CC

155

we twice suggested establishing contact with Mugabe, but both times we received in reply a categorical instruction not to interfere in someone else's business.

These exchanges revealed that on matters of support to revolutionary movements, the role of the ID was decisive.[83] However, both the ID and Ministry of Foreign Affairs had incentives to collaborate with each other. The ID was responsible for preparing speeches, background papers, and policy recommendations for the Central Committee of the Communist Party and often relied on data and information gathered by the KGB, the Ministry of Defense and Ministry of Foreign Affairs, and other relevant agencies.[84] However, its staff in Moscow numbered only 150 and its foreign network was quite limited. Therefore, it had an incentive to cooperate with the Ministry of Foreign Affairs which, through its role in staffing personnel in Soviet embassies throughout the world, had a much larger (and independent) capacity for gathering foreign intelligence.[85] In turn, collaboration with the ID afforded the Foreign Ministry with much greater access to influential party organizations.[86,i]

The Soviets also sponsored a variety of organizations and conferences that served to launder their role in providing aid to ZAPU and other African liberation movements. For instance, the Afro-Asian People's Solidarity Organization (AAPSO) was founded in Cairo in 1957 with extensive Soviet assistance,[88] and the Moscow branch of AAPSO, the Soviet Afro-Asian Solidarity Committee,[89] played an important role in managing the relationship with top ZAPU officials and funneling aid to the group. For instance, in a January 1961 meeting in Moscow with Tarcissius George (T. G.) Silundika, then a prominent leader of the National Democratic Party (NDP), the forerunner to ZAPU, the committee received a request from Silundika for various forms of aid, including funds for a printing shop and transportation and financial aid for leading members of the NDP.[90] This meeting was followed by

---

[i] One example of collaboration between the two organizations was a trip Soviet officials took in early 1967 to Tanzania, Zambia, Congo (Brazzaville), and Guinea. The members of the traveling party included Petr Manchkha, the head of the ID's African Section; Gennady Fomin, the head of one of the African Departments at the Ministry of Foreign Affairs; and Vadim Kirpichenko, who managed African affairs at the KGB. The trip was at the behest of the Central Committee of the Communist Party and impacted decisions taken at the Politburo regarding support to militant nationalists in Portuguese colonies in Africa.[87]

a July 1962 meeting of Joshua Nkomo with the Solidarity Committee during which Nkomo requested arms, explosives, and money to carry out sabotage and other activities in support of an armed uprising in Rhodesia.[91] In December 1963, James Chikerema, the vice president of ZAPU, conveyed a subsequent request to the Solidarity Committee for training in subversive and sabotage activities and in the manufacturing of small arms.[92] Officials from the ID and the Ministry of Foreign Affairs would often attend these meetings.

More generally, the conferences organized by AAPSO provided the Soviets with a means to gather information on the leaders of African liberation movements. For instance, before the January 1961 meeting with Silundika, the Soviet delegation at a November 1960 AAPSO Executive Committee meeting in Beirut reported that the NDP leader was "a modest and purposeful man, committed to his cause. He willingly gets in contact with Soviet representatives, and appears to be sincere with them."[93] Additionally, the AAPSO conferences also provided a forum to coordinate relations with various African liberation movements. In January 1969, AAPSO and the World Peace Council, another Soviet-backed organization, organized the International Conference of Solidarity with the Peoples of Southern Africa and the Portuguese Colonies in Khartoum, Sudan. Leaders from seven leading African liberation movements, including ZAPU, were invited to attend, and these groups came to be known as the "authentics" because they received official recognition from the OAU.[94,j] "Nonauthentic" groups more aligned with China boycotted the meeting, claiming the conference represented Moscow's attempt to exercise increasing control over the liberation struggle in Africa,[95] and these groups subsequently strengthened their ties to China.[96,k]

---

j   Other authentics included FRELIMO, SWAPO, the ANC of South Africa, the MPLA, PAIGC (Partido Africano da Independência da Guiné e Cabo Verde – African Party for the Independence of Guinea and Cape Verde), and the National Liberation Movement of the Comoros.

k   Several years earlier ZAPU, the ANC of South Africa, and other established groups launched a diplomatic initiative to delegitimize splinter groups and prevent them from gaining support. As a result of this effort, ZAPU blocked ZANU's attempt to join AAPSO by labeling the group as "pro-Beijing extremists." In general, at times ZANU's status as a nonauthentic impacted its relationship with authentics. In the late 1960s when FRELIMO consolidated its position in Tete Province in Mozambique, it initially looked to ZAPU, another authentic, to infiltrate northeastern Rhodesia through Tete. For FRELIMO's leaders, the liberation of Mozambique would be incomplete with Rhodesia under minority rule. Yet because of internal dissension, ZAPU was unable to collaborate with FRELIMO

Forms of Support

The Soviet Union was able to leverage a number of facilities within its territory to provide support to various African revolutionaries. Shubin noted that ZIPRA personnel were trained at a facility known as the "Northern Training Center" and also at the 165th Training Center at Perevalnoye in the Crimea.[98] Furthermore, he indicated that recruits from Zimbabwe received political training at the Institute of Social Sciences[99,1] and that ZAPU recruits also trained at the Air Force Center in Frunze in the Soviet Republic of Kirghizia.[101]

Two groups of ZAPU personnel were enrolled in mid-1964 into the Northern Training Center to commence a ten-month course that covered general military subjects and which specialized in guerrilla and conventional warfare.[102] Mavhunga also noted that later groups underwent twelve to eighteen months of instruction that consisted of communist indoctrination at the Central Komsomol (Communist Union of Youth) School in Moscow, as well as officer cadet training at the Odessa Military Academy in Ukraine. Subsequent training also occurred in Tashkent (in Uzbekistan) and Perevalnoye, where emphasis was placed on mobile warfare with artillery, airpower, heavy armor, speed, surprise, and "an orderly war theater delineated into 'tactical areas of responsibility'."[103]

As part of an oral history project, Alexander and McGregor interviewed seven ZIPRA intelligence personnel who underwent training in the Soviet Union. In speaking of the training of one such individual, Lazarus Ncube, in the Soviet Union, the authors noted:

> 'The most useful aspect was combat warfare, and we were trained in mechanical and chemical warfare.' He recalled his intelligence training with great pride. It had included, 'collecting combat information by stealth, spying on the enemy, numbers, equipment,

---

at this time, so FRELIMO worked with ZANU in Tete, allowing ZANU to acquire experience in guerrilla warfare, which it was eventually able to use against Rhodesia.[97]

[1] Shultz noted that the Institute of Social Sciences functioned as a center of instruction for communists from nonruling parties, and the curriculum included instruction on various operational methods for the illegal seizure of power. Instructors were provided by the KGB and GRU (Soviet military intelligence) and courses covered Marxism-Leninism, legal party building, political economy, and strategy and tactics of the world communist movement.[100]

identifying aircraft, artillery, tanks. So we were specialists in equipment, routine, notes for attack. We really used that on return. Also analyzing combat – making notes of events or sounds and submitting a report – I was good at that'. Stool Matiwaza felt that his Soviet training had allowed him to make a real contribution to the struggle: 'we were taught up to brigade level ... so we could come and teach other young soldiers how to be soldiers – that was the most important.'[104]

The accounts of trainees from other African countries provide interesting insights into the training offered in the Soviet Union. One member of *Umkhonto we Sizwe* (Spear of the Nation in Zulu), the military wing of the ANC of South Africa, noted:[105]

> We had undergone a course in the Soviet Union on the principles of forming an underground movement. That was our training; the formation of the underground movement, then building guerrilla detachments. The Soviets put a lot of emphasis on the building of these underground structures, comprising at the beginning very few people.

Another recruit noted:[106]

> We were taught military strategy and tactics, topography, drilling, use of firearms and guerrilla warfare. We also covered politics, with heavy emphasis on skills needed [for] construction and the use of explosives, vehicle maintenance, feeding a mobile army and first aid in the field: everything necessary for survival under guerrilla conditions.

Various recruits who trained in the Soviet Union went on to serve in influential positions in their home countries. General Siphiwe Nyanda, who eventually rose to become the chief of the South African National Defense Force, described his experience:

> In the USSR, we were staying in an apartment on Gorki Street, Moscow, where the lectures were conducted. For practical exercises, we went to a place outside Moscow. . . . We studied MCW (Military and Combat Work) as part of an abridged Brigade Commanders' course.

The course covered the following subjects, among others,

1. Communications

2. Underground work

3. Surveillance

4. Secret writing

5. Secret meetings

6. Photography

7. Military work

8. Ambush

9. Attack

10. Artillery effectiveness

11. Small arms

All were useful.[107]

Additionally, a number of important themes emerged from the interviews conducted by Alexander and McGregor, including an appreciation of Soviet military technology and the willingness of Soviet instructors to modify training to reflect conditions in Rhodesia. Following his arrival in Moscow in 1964, Dabengwa noted:[108]

> The training director . . . gave us the full description of what our course would comprise of. He wanted to know if there was anything that they had left out that we think we needed to be trained in. So we had those discussions and finally came up with a full training program which involved some of the things we'd mentioned.

In general, ZAPU trainees placed importance on adapting training and technology to local conditions, as noted by Alexander and McGregor:[109]

> The ZAPU cadres also consistently stressed the importance of adapting Soviet training and technology to the challenges of the Rhodesian war, which differed both from other southern African contexts and from the conditions faced by Soviet partisans in the Second World War. Weapons, for example, needed to be modified to suite the southern African climate. But the adaptations were more far-reaching than this. [Abel] Mazinyane explained that 'the ideas' needed to be adjusted to take account of Rhodesia's developed infrastructure and the dominance of Rhodesian air power. ....In contrast to the Soviet experience, ZIPRA lacked an air force, so, although 'the partisans could do some air drops for their supplies, with us we had to carry ours.' The collection of intelligence was also weighted towards 'working with the population' rather than the use of, for example, sophisticated listening devices.

Notably, the Soviet experience during the World War II, known as the "Great Patriotic War" to Russians, loomed large in the impression left upon ZIPRA trainees in the Soviet Union. Alexander and McGregor noted:[110]

> Soviet history, most importantly in the form of the Great Patriotic War, constituted a significant part of our interviewees' accounts of their political education. Soviet instructors created a powerful sense of the suffering and sacrifice of the Second World War through the use of films and visits to memorials. This history was also dramatically embodied by those instructors who were Second World War veterans of guerrilla war. As Shubin notes, the head of the 'Northern Training Centre' was for many years Major-General Fyodor Fedorenko, 'an ex-Second World War guerrilla commander in the Crimea.' For Dabengwa, ideological training was 'a history of the USSR really. They came in and .. . showed us how they had finally got to the stage where

they formed the Communist Party of the Soviet Union right down from Stalin and how they operated during the various wars they went through.' He was inspired by films about the Second World War and the personal stories of the partisans who taught his group. Dabengwa found the films 'of how the Russian army fought against the Germans . . . very moving': 'the dedication that they had – we were quite impressed. It gave us an urge to come back home and be able to do the same.'

These sentiments were echoed by Abel Mazinyane, whose training took place in the early 1970s:[111]

> [T]hey'd show you some films of the Soviet Union. It lost 20 million. Huge. But the German army was about how many kilometers from Moscow? Just close to the airport, as you enter the suburbs, there's a place where the Germans were stopped. It's called Volkalas. They were stopped there, then from there they drove them up to Berlin. They portray this patriotism – it touches you. People sacrificed so much. Some people stopping a machine gun, running into it to stop it so that others could advance. That had an impact on us who were fighting for our independence.

Another lasting impression was the experience of "living socialism," in particular the contrast between the relative egalitarianism of Soviet society with the discrimination, racism, and deep inequality recruits experienced in Rhodesia.[m] According to Alexander and McGregor:[113]

---

[m] Nonetheless, in the 1960s and 1970s, black Africans were at times regarded as exotic oddities by the Soviet population, and sometimes initial reactions consisted of simian ascriptions. Mazinyane noted "you're coming from a country that is predominantly black, then you go there, you're not used to being a minority. You get into a metro, an underground tube, everybody is looking at you. Then in Moscow, there were people, they'd never seen a black person. These people would do certain things, you'd feel it was racialism, but these people were curious. They'd want to touch your skin, to touch your hair, to ask how you live in Africa, all those things." Moyo noted that "the reactions towards us, we could understand. . . . We knew there were not blacks in the Soviet Union, unlike America and the UK, where our forefathers were taken as slaves. . . . So we were not even offended if we found people who wanted to see if we were human beings. . . . Some thought we were baboons . . . but when they heard us greeting them in Russian they realized we were human beings."[112]

By far the most compelling political lessons remembered by our interviewees were those that they derived from their encounters with 'living socialism,' and specifically the stark contrast they saw between Soviet egalitarianism and provision for citizens' needs and Rhodesia's deeply unequal and discriminatory society. Ncube remembers his shock at the very idea of human beings being equal: 'we couldn't believe it when they said you are equal.' He continued 'we wanted a new way of life where everyone was equal. Capitalism was the worst thing you could imagine.' The ZAPU cadres were astounded by what they saw as access for all to services, work and basic care. . . . The contrast with the Rhodesian state was profound: at home, Mazinyane explained, 'our group was not part of anything, so the state was suppressing us. There [in the USSR] was a state saying everyone was equal – that was good. That was the main message: this state was good, it gives everyone equal opportunity.' The idea that the state might work to create equal opportunities for all, rather than doing precisely the opposite, reinforced ZAPU cadres' view that they were fighting a 'system,' not a race.

As previously noted, Soviet training also took place in Africa. In July 1977, Soviet advisors arrived at a ZAPU camp near Luena, Angola, located close to the border with Zambia, and over the next year they trained more than ten thousand ZAPU members in military tactics, including guerrilla warfare.[114,n]

Additionally, at times the Soviets staffed personnel in embassies of surrounding countries to assist with political and military training. In July 1978, three Soviet advisors were sent to Lusaka, Zambia, to provide support to the political leadership of ZAPU and assist ZIPRA with planning and organizing combat in Rhodesia.[116] As indicated by Vassily Solodovnikov, the Soviet ambassador to Zambia:

> Outwardly, for the public, the group was assigned to Zambia's Ministry of Defense, but it didn't work even

---

n   Additionally, there are some indications that Soviet instructors supervised the training of ZAPU guerrillas in camps in Ghana in 1961–1962.[115]

a single day there. In reality, the military specialists worked as councilors to the Chief Commander of the People's Revolutionary Army, Joshua Nkomo. These people were first-class specialists in guerrilla warfare.[117]

The USSR likely provided a substantial amount of military aid to ZIPRA, although open-source documentation is somewhat unclear on the specifics of this assistance. Kempton noted that in the late 1970s large stockpiles of Soviet weapons destined for ZAPU were maintained in Zambia, and that at this time ZIPRA had Soviet-made SAM-7 systems.[118] Additionally, he noted that ZAPU possessed rocket-propelled grenade launchers, recoilless rifles, artillery up to 120 mm, and armored vehicles and advanced communications equipment, but he does not indicate whether they were of Soviet origin.[119] Additionally, he noted that Soviet arms were delivered by Cuban forces in Angola. Furthermore, Shubin noted that Soviet tanks and other heavy arms were provided to ZAPU but were not put to use given the Lancaster House Conference, and the presence of such arms may have made the Rhodesian government more willing to compromise.[120]

The Soviets also provided other forms of military assistance. Kempton noted that Soviet-trained intelligence agents led the infiltration of ZIPRA forces into Rhodesia, and that the USSR provided military radios in addition to building and running the main communications network at ZIPRA headquarters.[121] The Soviets also supported joint operations launched by ZAPU and the South African ANC in 1967–1968[122] and advised ZIPRA on logistics, intelligence, operations, communications, reconnaissance, and military organization.[123]

Other forms of Soviet support to ZAPU included advice from Soviet lawyers and diplomats during the Lancaster House Conference and talks in Geneva, airline tickets to international conferences,[124] and beginning in 1974 daily airtime on Radio Moscow for Africa (in English) for ZAPU broadcasts.[125,o]

---

[o] Before 1974 ZAPU had been able (since 1968) to intermittently use Radio Moscow transmitters for broadcasts.[126] Additionally, beginning in February 1979 Radio Moscow also began airing Nkomo's statements beamed toward Rhodesia, playing them in Shona and Ndebele.[127]

PEOPLE'S REPUBLIC OF CHINA

Actors and Methods Used to Provide Support

Various organizations in China were responsible for formulating and implementing policy on Africa. First and foremost, important decisions regarding Africa were made by the leadership of the Chinese Communist Party (CCP), specifically by the Politburo. Writing in the early 1970s, Larkin noted that "Most decisive decisions are probably made by members of the Standing Committee of the Political Bureau of the Central Committee, that handful of men—seven prior to the cultural revolution—at the apex of the CCP structure."[128]

In 1960 the Chinese foreign policy bureaucracy began to evolve to accommodate the achievement of independence by a number of African countries. That year the Chinese Ministry of Foreign Affairs established an African division to handle formal relations with the seventeen African nations that achieved independence in 1960, and the CCP Central Committee established a Special Committee in Charge of African Affairs, as well as the China African People's Friendship Association (CAPFA), which served as an umbrella organization for a number of "people's" organizations in China, including the All-China Federation of Trade Unions, the Women's Federation of the PRC, the China Peace Committee, the All-China Journalists' Association, and the All-China Students' Federation.[129] These various organizations under CAPFA invited various African groups to visit China, and their role in laundering the involvement of the CCP is well captured by Ogunsanwo:[130]

> In the field of propaganda and agitation, Marxist-Leninists have always found "people's" organizations indispensable. These can perform adequately the necessary activities without incurring the odium normally attached to Communist parties in many countries. No doubt many such organizations created to further Communist causes support noble and commendable issues. . . . Such causes as "peace," "disarmament," democratic freedoms and "women's rights" appeal to well-meaning individuals and groups who would normally ignore campaigns from Communist parties. It has been a logical step, therefore, for Communists to set up or sponsor the creation of organizations which

165

behind a façade of altruism can act as transmission belts for Communist ideas and aims. These organizations are of major importance where communist parties are non-existent, newly-founded or prohibited and where open canvassing is dangerous.

It should be noted that eight of the founding members of CAPFA were associated with the China Young Communist League, which was the youth arm of the CCP.[131]

AAPSO was another institutional mechanism that facilitated (as in the case of the Soviet Union) Chinese engagement with Africa. The key executive bodies within AAPSO consisted of the permanent secretariat (based in Cairo), the executive committee, the fund committee, and the control committee, and AAPSO enabled China to establish relationships with representatives of various national liberation movements in Africa.[132,p] Additionally, AAPSO, along with the OAU's Liberation Committee, may have facilitated Chinese financial and military support to various liberation movements in Africa. Jackson suggested that Chinese funding, training, and arms may have been funneled through AAPSO's Afro-Asian Solidarity Fund or the OAU's Liberation Committee,[134] while Chau indicated that the Liberation Committee oversaw guerrilla training camps in Tanzania and Zambia and distributed Chinese arms and ammunition.[135]

Although the details of China's financial and military support to various African movements through these two institutions remains somewhat opaque, it is clear that AAPSO served as a forum for the elaboration of the Sino-Soviet dispute. As noted earlier, during the 20th CPSU Congress in 1956, in addition to denouncing Stalin, the Soviet Union endorsed the "three peacefuls," specifically peaceful competition with the West, peaceful coexistence with the West, and the peaceful transition from bourgeois parliamentary democracy to socialism.[136] Soviet representatives at AAPSO subsequently attempted to convince the organization to adopt "peace" and "disarmament" as the organization's main goals,[137] and they achieved success in 1963 after an AAPSO meeting in Nicosia, in which the organization expressed its "loyalty to the principle of peaceful coexistence between independent states with

---

p   The support provided to African political organizations through AAPSO was managed by the All-China Afro-Asian Solidarity Committee, which was established in Beijing in 1958.[133]

different social and political systems . . . in the struggle for peace, complete and universal disarmament and the banning of nuclear tests and the liquidation of military bases."[138]

Chinese representatives at AAPSO vehemently opposed this stance, taking a more radical position by arguing that imperialism must be defeated before peace can be achieved. As one Chinese representative noted during a 1964 AAPSO meeting in Algiers:[139]

> A certain outside force . . . has been trying to impose on us an erroneous line which leaves out anti-imperialism and revolution. It spreads the nonsense that Afro-Asian peoples' task of opposing imperialism and old and new colonialism has been completed. . . . It propagates the view that the main task now confronting the Afro-Asian peoples in their struggle is "peaceful coexistence" with imperialism and old and new colonialism, and general and complete disarmament. This erroneous line in fact meant that the oppressed nations must forever suffer imperialist plundering and enslavement.

The reference to an "outside force" is notable because Chinese representatives attempted to have the Soviet Union expelled from AAPSO based on geographic and racial criteria. In a December 1961 meeting of the AAPSO Executive Council, China argued that the Soviet Union was not an Asian country and specifically that its Asian republics were part of a "European" political entity.[140] Additionally, the Chinese argued that the Soviet Union was a "white man's country" and therefore ill-suited to participate in an organization that represented oppressed "colored" races.[141] In a 1963 open letter to all party organizations, the Central Committee of the CPSU noted:[142]

> Beginning with the close of 1961, Chinese representatives in international democratic organizations have been openly imposing their erroneous views. . . . They opposed [the] participation of representatives of the Afro-Asian Solidarity Committees of the European Socialist countries in the third Afro-Asian People's Solidarity Conference in Moshi. The leader of the Chinese delegation told the Soviet representatives that "whites have no business here." At the journalists'

conference in Djakarta, the Chinese representatives followed a line designed to deprive Soviet journalists full-fledged delegate status on the plea that the Soviet Union . . . is not an Asian country.

Another important institution the Chinese used to establish contacts with African liberation movements was the New China News Agency (NCNA). For instance, in 1961 Ndabaningi Sithole, who led the ZANU split from ZAPU in 1963, met with a representative of the NCNA in Moshi, Tanzania.[143] In general, in addition to its role as "the primary PRC vehicle for the collection and dissemination of news at home and abroad,"[144] the NCNA had a prominent intelligence collection role,[145] and "NCNA correspondents overseas, with the agency's foreign offices as headquarters, [were] conducting activities of cultural infiltration, political united front [activities], and subversion."[146]

Forms of Support

As previously indicated, the main form of support the Chinese provided ZANU was military training, which was conducted in a variety of bases in Africa and in China. In July 1960 China and Ghana established diplomatic relations, and that same year China (along with the Soviet Union) helped establish four "freedom fighter" camps operated by Chinese instructors to train foreign fighters in guerrilla warfare tactics in preparation for operations in Rhodesia, South Africa, and the Portuguese colonies in Africa.[147] Additionally, in 1964 five Chinese guerrilla warfare experts arrived at a training facility in Half Assini, close to the border with the Ivory Coast, and they developed a twenty-day course on the use of explosives, guerrilla tactics, and "basic guiding and thinking on armed struggle."[148] The facility was eventually relocated to Obenemasi, and by January 1965, it featured 210 students and seventeen Chinese instructors.[149,q] Training included instruction in explosives, weapons, guerrilla warfare, and the use of telecommunications

---

q   Chinese instructors also collaborated with the Bureau of African Affairs, a nongovernmental organization Nkrumah established to coordinate aid to liberation movements elsewhere in Africa. The Bureau of African Affairs maintained a special bureau that handled militants, activities, and spies in other African countries, and it recommended that militants be trained by Chinese instructors and then sent on operations throughout Africa.[150] One analyst noted, however, that the Bureau of African Affairs became the headquarters for all "non-diplomatic" Ghanaian involvements in Africa.[151]

equipment, and students throughout Africa participated.[r] Chinese training camps were also located in Tanzania, specifically in Dar es Salaam, Bagamoyo, Moshi, Mgulani, Songea, Kongwa, Morogoro, Nachingwea, and Mtoni.[153,s]

Beginning in 1963, the training of Africans also occurred at three different facilities in China: one in Harbin, in Manchuria; another at Nanjing, on the Yangtze River; and a third at a location in Shantung Province, along the north coast.[155] The first group of ZANLA recruits trained in China included Emmerson Mnangagwa, who replaced Robert Mugabe as leader of Zimbabwe in late December 2017, as well as John Shoniwa, Eddison Shirihuru, Jameson Mudavanhu, and Lawrence Swoswe.[156] In addition to military science, they also received ideological instruction, including listening to lectures with titles "the Chinese Revolutionary Struggle," "the People's War," and "the Democratic Revolution."[157]

A number of ZANLA personnel were trained in Nanjing, and in addition to physical training, ZANU recruits were instructed in the theory of guerrilla war, sabotage, and the use of heavy arms, such as machine guns, bazookas, and antitank mines.[158] Recruits also received training in making simple bombs and instructed on how to employ them to sabotage railway lines, supply depots, and military and police outposts.[159] Training lasted between six and nine months and included instruction on the works of Mao as well as basic communist pedagogy so that recruits understood that communism was similar to African communal life prior to the arrival of Europeans.[160] The training of African recruits appears to have been conducted by the International Equipment Division of the General Rear Services Department of the Chinese People's Liberation Army.[161]

Mao is forever remembered as one of the pre-eminent theorists of guerrilla warfare, having likened guerrillas as the fish that swim

---

[r]   Chinese training of African recruits in Ghana ended in 1966 after Nkrumah was ousted through a military coup. The new Ghanaian authorities sought to substantially reduce the influence of Communist countries in Ghana. More than 1,000 personnel from the Soviet Union, China, and Eastern Europe were expelled, and 430 Chinese nationals, including three intelligence officers and 13 guerrilla warfare specialists, were forced to leave Ghana. Diplomatic relations between the two countries were suspended in 1966, and they did not resume until 1972.[152]

[s]   The operation of the camps in Tanzania and Ghana likely involved the International Liaison Department of the CCP, which supported revolutionary groups throughout Africa and Asia.[154]

through the sea of the population. More specifically, in *On Guerrilla Warfare*,[1] he noted that:[163]

> Because guerrilla warfare basically derives from the masses and is supported by them, it can neither exist nor flourish if it separates itself from their sympathies and cooperation. . . . Many people think it is impossible for guerrillas to exist for long in the enemy's rear. Such a belief reveals [a] lack of comprehension of the relationship that should exist between the people and the troops. The former may be likened to water and the latter to the fish who inhabit it.

Hence, one noteworthy aspect of the instruction imparted by the Chinese was training in how to politicize the populace. At Nanjing, recruits were given instruction in mass mobilization, which had a big impact on Josiah Tongogara, who was instrumental in changing ZANLA's tactics in the early 1970s to favor the politicization and mobilization of the populace. Speaking of the training he imparted on ZANLA recruits after he returned from China, he noted that "I had trained them in generalized guerrilla warfare and specialized mass mobilization."[164] Additionally, the training imparted on ZANLA recruits in China covering mass mobilization and politicization had a "hands-on" component, as noted by one ZANLA recruit:

> At the same time we practiced trying to politicize each other, imagining that the other chap was a Rhodesian peasant. This made us laugh a lot, but the Chinese instructor took it very seriously.[165]

It was this form of instruction, however, that appears to have made a difference in the conflict. Taylor noted:

> The involvement of China with ZANU had, however, a profound effect on the course of the liberation war in Rhodesia . . . ZANLA's military tactics underwent

---

[1] Although *On Guerrilla Warfare* was published in 1937, Mao started politicizing Chinese peasants much earlier. He started organizing peasant organizations in 1925, and he warned peasants of the threat imperialism posed toward countries in Asia and Africa. By mid-1926, Mao organized twenty "Associations to Erase Humiliation," where the slogan "Strike against Great Powers, Wipe Clean the National Humiliation" was employed. Additionally, around this time, he became principal of the Peasant Movement Training Institute, where he was responsible for mobilizing peasants in the countryside.[162]

a transformation from conventional military tactics, to the Maoist model of mass mobilization of the population. This cardinal principle of Maoist military thought was rigidly adhered to by ZANLA throughout the Second Chimurenga. This gave ZANU a far firmer base in the Rhodesian countryside and a stronger support than their political rivals ZAPU. The process was gradual and was masterminded by the commander of ZANLA, Josiah Tongogara who, after his training in China, emphasized "guerrilla warfare and specialized mass mobilization."[166]

The change in tactics was brought about by the failure of a number of separate ZAPU and ZANU incursions into Rhodesia, and its importance was attested by various ZANU political and military leaders. Herbert Chitepo, the national chairman of ZANU, indicated that ZANU "tried to correct [ZAPU's errors] by politicizing and mobilizing the people before mounting any attacks. . . . After politicizing our people, it became easier for them to cooperate with us and identify with our program and objective."[167] Mugabe noted that "there was a complete revision of our manner of carrying out the armed struggle. We began to realize that the armed struggle must be based on the support of the people. . . . We worked with Frelimo for eighteen months in Tete province [in Mozambique]. It was there that we learned the true meaning of guerrilla war."[168,u]

The emphasis on the direct mobilization of the populace appears to have been a distinguishing feature of Chinese training when compared with the instruction received in the Soviet Union. Rex Nhongo, who became ZANLA commander after the death of Tongogara and who had trained in the Soviet Union as a member of ZAPU, noted that "in the Soviet Union they had told us that the decisive factor of the war is the weapons . . . Chinese instructors [said] that the decisive factor was the people. This was a contradiction."[170]

There are various accounts of ZANU guerrillas' efforts to politicize the populace. One resident of Dande noted:

---

u   Recruits from FRELIMO were also trained by the Chinese. About fifty Mozambicans were trained in China and others were trained in Tanzania, and Taylor noted that FRELIMO, like ZANU, adopted Maoist guerrilla warfare tactics.[169]

171

They began to teach us politics. They said: We are called comrades not terrorists. We came from Mozambique because the government of Rhodesia does not treat us as equals. A white man works for a month and has enough money to buy a car. Can you people do the same as he can? We answered no.[171]

One former guerrilla gave the following account of a *pungwe*:

When we came to a village, the first thing we would do is hold a rally. The Commander and the Political Commissar would go to the place where the rally was to be held. The rest of us would go to all the houses and make sure that everyone came. Some people would want to come, others would not but there was no choice about it. You had to come. We wanted everyone there because if anyone wasn't they could go off quietly and betray us.

First we would explain who we were. We were ZANLA. We were not ZANU. ZANU was a political organization. We were the military wing. It was ZANU's job to go to other countries, to talk and negotiate. We did not go to other countries. We did not need the help of solders from other countries. We were Zimbabweans in Zimbabwe. And we did not use talk, we used guns.

We explained the structures of the ZANU party. We then explained the structure of the army and told the masses about the army high command and described their duties. Next we explained national grievances,[v] then colonialism, then neo-colonialism and capitalism. We explained that ours would be a socialist government and what that would mean to the masses.

The pattern of the meeting would be: talk for half an hour, then teaching the masses songs for an hour, then

---

[v]   In addition to being educated on the works of Mao, ZANLA recruits were educated on what were known as the National Grievances, which consisted of the main social, political, and economic grievances the black population experienced under the white minority government. These grievances included the dispossession of land, inequalities in education and health, political oppression, and low wages.[172]

talking for another half hour and so on, so that the people did not get bored.

While this rally was taking place, one of us would go on to the next village and quietly find out how much support we had there, whether it was safe for us to enter. He would just look like an ordinary person. There was no way you could tell he was a comrade.

At the end of the meeting we would say to the older people: "Mothers and Fathers, go home now and sleep in peace. But the children you must stay here." The younger people would stay, and we would then say: "What is our support here? Are people in favor of us, are people speaking against us and who is doing so?" Then they would tell us, for example, that some people were saying that they didn't have enough food to eat themselves without giving some to us. And many other complaints came out as well.[173]

At some meetings with the peasantry, ZANU guerrillas emphasized immediate concerns rather than abstract notions related to colonialism and capitalism, as indicated in another account by a ZANU guerrilla:

We would then get into an area, study the problems in that particular area, and then teach those people about their problems, how we can solve them by fighting the enemy. . . . Overall the land question was our major political weapon. The people responded to it. As for socialism versus capitalism, since the olden days of our ancestors our people used to work communally and live communally, which was almost the same as socialism.[174]

Hence, these accounts reveal the influence of Chinese methods of mass mobilization on ZANU operations. These practices, and in particular the direct mobilization of the populace by guerrilla forces through *pungwes*, appears also to have been a determining factor in the 1980 election that brought Mugabe to power. As one ZAPU official noted:

The PF [Patriotic Front] lost the election two years ago when ZANU began intensive political campaigning, using ZANLA to politicize the masses. ZANLA moved into the former ZIPRA areas at this time, such as Mashonaland West. They held pungwes for two years covering 80 percent of the country. This is the most important factor about the election results.[175,w]

However, some aspects of the instruction imparted by the Chinese were not well received by some ZANU commanders. One ZANU recruit who was sent to China for training noted:

> It was interesting that before we left Africa a ZANU commander had told us to work hard on the military tactics we would learn in China. "But," he added, "don't worry about politics—the Chinese have funny ideas about this. Don't listen to them, be careful, because they will try and make you think like they want you to think."[177]

This caution was apparently in reference to anti-elitist sentiment and attitudes conveyed by Chinese instructors during training. This same recruit noted:

> And when he [the Chinese instructor] told us about class struggle, I learnt something else the ZANU officials would not like to hear—that there is no point fighting to get a white elite out, just to have a black elite take over. I realized that, if this was going to be the case, we were fighting for nothing. Why should we be risking our lives to remove Smith, the white dictator, to put some black dictator in his place? This is why politics is so important. If a soldier knows he is fighting for a better life then he is prepared to die for this.

> Our instructor told us that the revolution would never be achieved if it was based on a lie. It is no use to say to the people, "Look, the whites are rich and powerful. If you remove them, you will be rich and powerful," and talk about self-government, independence and freedom. What you have to promise is to change

---

the entire political life. The people must understand what it means to fight imperialism and capitalism: they are fighting elitism and individual greed—they are fighting collectively, to share the riches of the country collectively. They must be told that unless the wealth of the country is shared, an elite will take over again, and they'll be back to where they were before independence. They'll be just as poor and just as discriminated against, not as blacks any more, but as an inferior class.[178]

He also stated:

We were also told in China that every good soldier must be free to criticize his leader and every leader must expect this. This was the only way to make sure everyone was clear all the time about how to fight. Discussion is very important for morale. It is very important to build up trust between leaders and men. Well, in Chunya, you saw how much criticism we were allowed to do![179]

Apparently, and somewhat ironically, for some ZANU commanders there were limits to the benefits of the Chinese emphasis on politicization and antielitism. That is, although it was seen as advantageous when it came to challenging the hierarchical authority relationship between the white minority government and the disenfranchised and economically marginalized black majority, it was seen as dangerous if it led ZANU recruits to challenge what some commanders saw as their own hierarchical relationship with recruits. When this particular recruit returned to Tanzania and questioned ZANU commanders regarding their plans, the commander did not appreciate such effrontery and eventually interrogated a number of soldiers in the camp, demanding from each an oath of allegiance to ZANU and to the freedom struggle in Rhodesia.[180] Nonetheless, Chinese ideas on guerrilla warfare assumed a canonical status among ZANLA forces. Moorcraft and McLaughlin noted that "by 1978 the works of Mao were ZANLA's bible of guerrilla warfare,"[181] and ZANLA guerrillas carried copies of Mao's Little Red Book with them for reference.[182] Mao's teachings were also incorporated into Chimurenga songs to motivate ZANLA guerrillas, as in the following:[183]

> *There are ways of Revolutionary soldiers in behaving*
> *Obey all orders*
> *Speak politely to the people*
> *We must not take things from our masses*
> *Return everything captured from the enemy*
> *Pay fairly for what you buy*
> *Don't take liberties with women, don't ill-treat captives of war*
> *Don't hit people too severely*
> *These are the words said by the people of ZANU teaching us*
> *These are the words said by Chairman Mao when teaching us*

NOTES

[1] Rhodesian Ministry of Foreign Affairs, *Communist Support and Assistance to Nationalist Political Groups in Rhodesia* (Rhodesia: Information Section, Ministry of Foreign Affairs, November 28, 1975), posted on *Rhodesia and South Africa: Military History* (blog), accessed August 22, 2014, http://www.rhodesia.nl/commsupp.htm.

[2] Vladimir Shubin, *The Hot 'Cold War': The USSR in Southern Africa* (London: Pluto Press, 2008), 179.

[3] Rhodesian Ministry of Foreign Affairs, *Communist Support and Assistance to Nationalist Political Groups in Rhodesia.*

[4] Daniel R. Kempton, *Soviet Strategy toward Southern Africa: The National Liberation Movement Connection* (New York: Praeger, 1989), 104.

[5] "Zimbabwe: Second Chimurenga, 1966-1979," March 28, 2015, http://www.worldhistory.biz/sundries/39661-zimbabwe-second-chimurenga-1966-1979.html, retrieved December 1, 2017

[6] Paresh Pandya, *Mao Tse-Tung and Chimurenga: An Investigation into ZANU's Strategies* (Johannesburg: Skotaville Educational Division, 1988), 48–49.

[7] Ibid., 47–48.

[8] Ibid., 48.

[9] Ibid., 105.

[10] Ibid., 49.

[11] Kempton, *Soviet Strategy toward Southern Africa*, 112.

[12] William Cyrus Reed, "International Politics and National Liberation: ZANU and the Politics of Contested Sovereignty," *African Studies Review* 36, no. 2 (1993): 44.

[13] Ibid.

[14] Shubin, *The Hot 'Cold War,'* 181–182.

15 Alaba Ogunsanwo, *China's Policy in Africa, 1958–71* (London: Cambridge University Press, 1974), 172.

16 Shubin, *The Hot 'Cold War,'* 165.

17 Clarence Chongo, "Decolonising Southern Africa: A History of Zambia's Role in Zimbabwe's Liberation Struggle, 1964-1979," (Ph.D. dissertation, University of Pretoria, 2015), 68.

18 Kempton, *Soviet Strategy*, 106.

19 Gilbert M. Khadiagala, *Allies in Adversity: The Frontline States in Southern African Security, 1975-1993* (Athens, Ohio: Ohio University Press, 1994), 28.

20 Pandya, *Mao Tse-Tung and Chimurenga*, 49.

21 Ibid.

22 O. Igho Natufe, *Soviet Policy in Africa: From Lenin to Brezhnev* (Bloomington, IN: iUniverse, Inc., 2011), 163.

23 Chongo, "Decolonising Southern Africa," 138.

24 Ibid.

25 Ibid., 141.

26 Ibid., 67.

27 Ibid., 75–76.

28 Ibid., 77.

29 Ibid.

30 Ibid.

31 Ibid., 78.

32 Ibid., 133.

33 Ibid., 83–84.

34 Ibid., 86, 89.

35 Ibid., 120.

36 Ibid., 92–93.

37 Ibid., 139.

38 Ibid., 140.

39 Pandya, *Mao Tse-Tung and Chimurenga*, 49.

40 Chongo, "Decolonising Southern Africa," 139.

41 Ibid., 335–336.

42 Ibid., 332–336.

43 Pandya, *Mao Tse-Tung and Chimurenga*, 50.

44 Ibid.

45 Chongo, "Decolonising Southern Africa," 169.

46 Vladimir Shubin, "Moscow and Zimbabwe's Liberation," *Journal of Southern African Studies* 43, no. 1 (2017): 230.

47 Ibid.

48 Ibid.

49 Clapper ton Chakanetsa Mavhunga, "A Plundering Tiger with its Deadly Cubs? The USSR and China as Weapons in the Engineering of a 'Zimbabwe Nation,' 1945-2009"

in *Entangled Geographies: Empire and Technopolitics in the Global Cold War* (Cambridge, MA: MIT Press, 2011), 237.

50 Wazha G. Morapedi, "The Dilemmas of Liberation in Southern Africa: The Case of Zimbabwean Liberation Movements and Botswana, 1960-1979," *Journal of Southern African Studies* 38, no. 1 (2012): 75.

51 Ibid.

52 Ibid.

53 Ibid.

54 Ibid., 76.

55 Ibid., 77.

56 Ibid., 78, 81, 83.

57 John Daniel, "Racism, the Cold War and South Africa's Regional Security Strategies 1948-1990", in ed. Sue Onslow, *Cold War in Southern Africa: White Power, Black Liberation* (London: Routledge, 2009), 37.

58 Ibid., 37.

59 Ibid., 39; and Richard Leonard, *South Africa at War: White Power and the Crisis in Southern Africa* (Westport, Connecticut: Lawrence Hill and Company, 1983), 83.

60 Annexure 1 to Document HSL/131/63/1 from archives of the South African Defense Forces (Declassified on April 19, 2011).

61 Leonard, *South Africa at War*, 83–84.

62 Ibid., 85.

63 Ibid., 83.

64 Pandya, *Mao Tse-Tung and Chimurenga*, 194.

65 Ibid., 51, 54, 87.

66 Ibid., 54.

67 Ibid., 87–88.

68 Ibid., 88.

69 Ibid.

70 Ibid., 59.

71 Ibid., 194.

72 Ibid., 58.

73 Ibid., 58–59.

74 Shubin, *The Hot "Cold War,'* 1, 7, 166.

75 Mark Kramer, "The Role of the CPSU International Department in Soviet Foreign Relations and National Security Policy," *Soviet Studies* 42, no. 3 (1990): 429.

76 Ibid., 430.

77 Shubin, "Moscow and Zimbabwe's Liberation," 229.

78 Shubin, *The Hot 'Cold War,'* 167.

79 Ibid., 168.

80 Kramer, "Role of the CPSU International Department," 431.

81 Quoted in Shubin, *The Hot 'Cold War,'* 185–186.

82 Ibid., 186.

[83] Ibid.

[84] Ibid., 432.

[85] Ibid.

[86] Ibid.

[87] Ibid., 167.

[88] Kempton, *Soviet Strategy*, 98.

[89] Richard H. Shultz, *The Soviet Union and Revolutionary Warfare: Principles, Practices, and Regional Comparisons* (Stanford: Hoover Institution Press, 1988), 80.

[90] Shubin, *The Hot 'Cold War,'* 152.

[91] Ibid., 153.

[92] Ibid., 154.

[93] Ibid., 152.

[94] Ian Taylor, *China and Africa: Engagement and Compromise* (London: Routledge, 2006), 29.

[95] Ibid.

[96] Kempton, *Soviet Strategy*, 103.

[97] Taylor, *China and Africa*, 113; Reed, "International Politics and National Liberation," 37, 41; and David Martin and Phyllis Johnson, *The Struggle for Zimbabwe* (London: Faber and Faber, 1981), 15–20.

[98] Shubin, "Moscow and Zimbabwe's Liberation," 231.

[99] Shubin, *The Hot 'Cold War,'* 155.

[100] Shultz, *The Soviet Union and Revolutionary Warfare*, 39.

[101] Shubin, *The Hot 'Cold War,'* 176.

[102] Ibid., 155; and Mavhunga, "A Plundering Tiger with its Deadly Cubs?" 239.

[103] Mavhunga, "A Plundering Tiger with its Deadly Cubs?" 240.

[104] Jocelyn Alexander and JoAnn McGregor, "African Soldiers in the USSR: Oral Histories of ZAPU Intelligence Cadres' Soviet Training, 1964-1979," *Journal of Southern African Studies* 43, no, 1 (2017): 58.

[105] Vladimir Shubin, "Unsung Heroes: The Soviet Military and the Liberation of Southern Africa," *Cold War History* 7, no. 2 (2007).

[106] Ibid.

[107] Quoted in Shubin, *The Hot 'Cold War,'* 255.

[108] Alexander and McGregor, "African Soldiers in the USSR," 56.

[109] Ibid.

[110] Ibid., 59.

[111] Ibid., 60.

[112] Ibid., 62–63.

[113] Ibid., 61–62.

[114] Shubin, *The Hot 'Cold War,'* 172–173.

[115] Kempton, *Soviet Strategy*, 98.

[116] Shubin, *The Hot 'Cold War,'* 173.

[117] Quoted in Shubin, *The Hot 'Cold War,'* 173.

[118] Kempton, *Soviet Strategy*, 106.

119 Ibid.

120 Shubin, *The Hot 'Cold War,'* 183.

121 Kempton, *Soviet Strategy,* 107.

122 Shubin, *The Hot 'Cold War,'* 156.

123 Kempton, *Soviet Strategy,* 107.

124 Shubin, *The Hot 'Cold War,'* 180.

125 Kempton, *Soviet Strategy,* 107.

126 Lebona Mosia, Charles Riddle, and Jim Zaffiro, "From Revolutionary to Regime Radio: Three Decades of Nationalist Broadcasting in Southern Africa," *African Media Review* 8, no. 1 (1994): 13.

127 Kempton, *Soviet Strategy,* 107.

128 Bruce D. Larkin, *China and Africa, 1949–1970* (Berkeley, CA: University of California Press, 1971), 214.

129 David H. Shinn and Joshua Eisenman, *China and Africa: A Century of Engagement* (Philadelphia: University of Pennsylvania Press, 2012), 61; and Ogunsanwo, *China's Policy in Africa,* 97.

130 Ogunsanwo, *China's Policy in Africa,* 41.

131 Shinn and Eisenman, *China and Africa,* 61.

132 Donovan C. Chau, *Exploiting Africa: The Influence of Maoist China in Algeria, Ghana and Tanzania* (Annapolis: Naval Institute Press, 2014), 30.

133 Shinn and Eisenman, *China and Africa,* 59.

134 Steven F. Jackson, "China's Third World Foreign Policy: The Case of Angola and Mozambique, 1961–93," *The China Quarterly,* no. 142 (1995): 393, fn. 18.

135 Chau, *Exploiting Africa,* 31.

136 O. Igho Natufe, *Soviet Policy in Africa: From Lenin to Brezhnev* (Bloomington, IN: iUniverse, Inc., 2011), 153.

137 Ogunsanwo, *China's Policy in Africa,* 93.

138 Natufe, *Soviet Policy in Africa,* 167.

139 Ibid., 167–168.

140 Ibid., 166.

141 Ibid.

142 Quoted in Natufe, *Soviet Policy in Africa,* 165.

143 Taylor, *China and Africa,* 107.

144 Jeffrey T. Richelson, *Foreign Intelligence Organizations* (Cambridge, MA: Ballinger Publishing Company, 1988), 293.

145 Chau, *Exploiting Africa,* 23,25.

146 Ibid., 25.

147 Ibid., 81.

148 Ibid., 89.

149 Ibid., 90.

150 Ibid., 79, 92.

151 W. Scott Thompson, *Ghana's Foreign Policy 1957–1966* (Princeton, NJ: Princeton University Press, 1969), xxiii.

[152] Chau, *Exploiting Africa*, 94–99.

[153] Ibid., 122, 139.

[154] Richelson, *Foreign Intelligence Organizations*, 278.

[155] Chau, *Exploiting Africa*, 90.

[156] Pandya, *Mao Tse-Tung and Chimurenga*, 28.

[157] Ibid., 27–28.

[158] Michael Raeburn, *We Are Everywhere: Narratives from Rhodesian Guerrillas* (New York: Random House, 1978), 45, 50.

[159] Pandya, *Mao Tse-Tung and Chimurenga*, 88–89.

[160] Ibid.

[161] Jackson, "China's Third World Foreign Policy," 393, fn. 18.

[162] Gregg Brazinsky, *Winning the Third World: Sino-American Rivalry During the Cold War* (Chapel Hill: University of North Carolina Press, 2017), 23.

[163] Mao Tse-Tung, *On Guerrilla Warfare*, translated by Samuel B. Griffith II (Urbana: University of Illinois Press, 2000), 44, 92–93, cited in Benjamin Valentino, Paul Huth and Dylan Balch-Lindsay, " 'Draining the Sea:' Mass Killing and Guerrilla Warfare," *International Organization* 58, no. 2 (April 2004): 384.

[164] Pandya, *Mao Tse-Tung and Chimurenga*, 89.

[165] Raeburn, *We Are Everywhere*, 50.

[166] Quoted in Taylor, *China and Africa*, 108–109.

[167] Ibid., 109.

[168] Quoted in David Lan, *Guns and Rain: Guerrillas and Spirit Mediums in Zimbabwe* (Berkeley, CA: University of California Press, 1985), 124.

[169] Taylor, *China and Africa*, 94.

[170] Quoted in Taylor, *China and Africa*, 110.

[171] Lan, *Guns and Rain*, 127.

[172] Paul L. Moorcraft and Peter McLaughlin, *The Rhodesian War: A Military History* (Mechanicsburg, PA: Stackpole Books, 2008), 75.

[173] Lan, *Guns and Rain*, 127–128.

[174] Ibid., 128.

[175] Martin and Johnson, *The Struggle for Zimbabwe*, 331–332.

[176] Ibid., 328–329.

[177] Raeburn, *We Are Everywhere*, 46.

[178] Ibid., 46, 49.

[179] Ibid., 49–50.

[180] Ibid., 56.

[181] Moorcraft and McLaughlin, *The Rhodesian War*, 74.

[182] Pandya, *Mao Tse-Tung and Chimurenga*, 104.

[183] Moorcraft and McLaughlin, *The Rhodesian War*, 101–102.

CHAPTER 6.
CONCLUSIONS

The Rhodesian case offers what social scientists would regard as a "natural experiment," as it allows for a comparison of the impact of two different approaches to guerrilla and unconventional warfare within a particular setting. Ultimately the efficacy of the support offered to Zimbabwe African National Union (ZANU) and Zimbabwe African People's Union (ZAPU) must be judged by whether the support helped achieve sponsor objectives, which was the dismantlement of the white minority regime and its replacement by a government based on majority rule and led by the recipient of aid. In this regard various analysts have concluded that the "people's war" approach favored by China, which found direct expression in the Zimbabwe African National Liberation Army (ZANLA) attempt at direct mobilization of the populace through *pungwes*, to have been decisive both in the conduct of the war against the minority regime in Salisbury and in the 1980 election.

From a sponsor perspective, China's support for Mugabe and ZANU during the liberation struggle paid long-term dividends. Writing in the late 1990s, Taylor noted that "The Republic of Zimbabwe has been the People's Republic of China (PRC's) most important link in southern Africa, though South Africa is now Beijing's major economic partner in the region and will likely supplant Zimbabwe in the future."[1] Bilateral relations appear set to remain solid even after Mugabe exited power in late 2017 following thirty-seven years as Zimbabwe's leader. His tenure was characterized by kleptocracy, authoritarianism, economic mismanagement, and the massacre of Ndebele civilians in Matabeleland in the mid-1980s. He was replaced by Emmerson Mnangagwa who, as noted in this report, received training in China in the early 1960s. In April 2018, Mnangagwa visited Beijing to reaffirm ties with the PRC, and the trip to the former sponsor was Mnangagwa's first overseas trip outside of Africa in the post-Mugabe era.[2]

Regarding the impact of ZANU's people's war strategy on the 1980 election, Gregory noted:

> The [1980] election campaign revealed, however, that ZIPRA [Zimbabwe People's Revolutionary Army] had not only fought in a smaller area of the country than ZANLA,[a] but that it had also adopted a different and,

---

[a] In contrast to the assertion that ZANLA fought in a larger area of the country, Brickhill noted that by the end of the war ZIPRA operated over as much territory as did ZANLA and was able to do so with fewer than half the number of guerrillas, since

in terms of mobilising popular support, a less success-ful guerrilla strategy. While ZIPRA had established an effective network of support for its guerrillas in their operational areas, it placed less emphasis than ZANLA on politicising the population or preparing people for a sustained and long drawn-out struggle. Despite the debates of the early 1970s and ZAPU's recogni-tion of its shortcomings, ZIPRA's war remained mili-tary rather than political in character. ZIPRA relied on scoring spectacular successes against the Security Forces or symbolic targets, such as the shooting down of two Viscount airplanes on civilian routes, in order to generate political support. The task of winning the population to ZAPU's political program was left to the network of supporters which had been built up throughout the country in the early 1960s when ZAPU operated as a mass-based party, and had gone under-ground when the party was banned. This division of labour between the guerrillas and the party cadres during the war was a central reason for the party's fail-ure at the polls. Many of the ZAPU underground struc-tures outside Matabeleland, in fact, collapsed during the war and thus the party was left organisationally weak in Mashonaland and struggling to compensate for the withdrawal of its operational arm, ZIPRA, to the assembly points.[4]

Gregory noted that ZAPU was able to convert support for ZANLA guerrillas into votes in the 1980 election through "teach-ins" that instructed the populace how to vote:[5]

It was this organisation of the peasantry in the rural areas during the war that provided the cornerstone for ZANU (PF)'s triumph at the polls. In the many areas in the eastern half of the country where the population had become an integral part of ZANLA's insurgency campaign, ZANU (PF)'s major objective during the election was to ensure that support for

---

ancillary tasks, such as the political mobilization of the populace and logistics, were han-dled by ZAPU rather than by ZIPRA.[3]

its guerrillas was translated into votes. Thus, having won a substantial proportion of the electorate to its cause during the war, ZANU (PF) did not set out, or indeed need, to convert voters or argue points from its manifesto. Rather it sought to ensure that its supporters knew when and how to vote, and that morale was maintained while most of the guerrillas had left for the Assembly Points under the terms of the ceasefire. An important feature of the ZANU (PF) election campaign in this context was its "teach-in" rallies. With an estimated minimum of 44 percent and maximum of 67 percent of the adult black population functionally illiterate, ZANU (PF) made a point of holding instruction courses on how to vote and concentrated on those areas of highest illiteracy.

The impact of ZANLA's mobilization of the populace was also noted by Ken Flower, the head of Rhodesia's Central Intelligence Organization (CIO). In his memoirs, he noted:

The disarray of ZANU and ZAPU in Zambia continued, and CIO continued to foment it. We also kept an eye on training camps in Tanzania and were aware that, following the arrival of Chinese instructors at the Itumbi camp in 1969, ZANLA too was undergoing a change of tactics along the lines of Mao Zedong's teaching—to politicize the masses before preparing to strike. However, we considered that as neither ZANU nor ZAPU had managed to mobilize support in areas adjacent to the Zambian border between 1964 and 1969 there was no reason why they should manage to do so now. The point we missed was that ZANU, in particular, would have a much better chance of mobilizing support in the north-eastern area bordering Mozambique because the area was far more densely populated and most of the population were Shona-speaking.[6,b]

---

[b] Other analysts have emphasized the aesthetics and form in which mobilization took place. Plastow noted, "Here, and in particular at the ZANU bases in Mozambique, increasing numbers of guerrillas were trained in the tactics of guerrilla warfare. They

Hence, it appears as if the Chinese emphasis on the politicization and mass mobilization of the population paid electoral dividends to ZANLA, whose reliance on a Maoist "people's war" strategy was successful in the Shona regions of the country. It is useful to recall the rationale for why ZAPU opted for a conventional strategy toward the end of the war. Writing in the mid-1990s, Brickhill noted:[8]

> No African liberation movement has actually seized power from a colonial regime. As a rule African independence is preceded by negotiations in which the colonial power attempts to restrain African nationalism. . . . Today, as Zimbabwe struggles to implement structural adjustment policies determined by international bankers and, a full decade after Independence, still seeks a land reform program to return land to the peasantry, the limitations of guerrilla warfare as a strategy for revolutionary transformation are more apparent than ever. ZIPRA's strategy, aimed at achieving a military victory, would certainly have created alternative options for transforming Zimbabwe after Independence.

---

were also trained in socialist theory and the importance of working with, and raising the political awareness of the masses within Rhodesia. From 1976 guerrilla incursions rapidly increased in number. Armed attacks became commonplace. But just as important in winning the war were the people's committees; these were set up in guerrilla areas, regularly held nocturnal *pungwes*—all-night rallies—sometimes several times a week in the villages, and led to a highly conscientised home base, especially in the largely Shona areas of ZANU influence. The politicization process called heavily upon the performance arts. In the guerrilla camps political discussion was carried out primarily through the medium of theatre. Traditional dances were revived and a huge number of liberation songs were carried into Rhodesia for consciousness-raising purposes. The *pungwes* made use of all these performance forms, both to revive a sense of cultural identity and to illustrate political messages. . . . From the specifically cultural viewpoint, what is noteworthy is how the process of involving the rural masses in the struggle by identifying popular mass culture with the liberation movement immensely strengthened the guerrillas position in areas they infiltrated. Because they spoke to people in language and terms relevant to their concerns, and used popular music and performance media to spread their messages, the guerrillas were able to subvert the ponderous machinery of colonial repression and propaganda. Political and cultural identification with the rural people gave ZANU its increasingly safe bases within the country, and the policy reaped its rewards when, in April 1980, ZANU won fifty-seven of the eighty seats available to black voters, so acquiring the right to set up the first independent government of a now renamed Zimbabwe."[7]

However, given that Operation Zero Hour was largely pre-empted by the Lancaster House talks and that ZANU obtained power using an alternative military strategy, perhaps a harsh rendering of ZAPU's choice of military strategy in the late 1970s was that it represented a strategy deprived of its Waterloo (although one can plausibly claim that the buildup of conventional forces by ZAPU influenced the peace talks at Lancaster House). However, it would be incorrect to say that ZAPU did not use guerrilla tactics or that it did not undertake population-based measures. Throughout the war ZIPRA guerrillas carried out mine warfare, sabotage operations, and raids and ambushes[9] but generally interacted with the population through ZAPU party officials. Brickhill noted:

> ZIPRA guerrillas related to the civilian population wherever possible through specific links to party branch structures. The commissars oversaw this relationship, but guerrilla logistics officers, security and intelligence officers and training instructors all sought to develop links with specific party officials responsible for a particular part of the relationship with civilians . . . it is clear that this approach by ZAPU and ZIPRA was mutually beneficial and supportive in a wide variety of ways. Crucially, many of the organizing, mobilizing and logistical tasks of the war could be carried out by the party, leaving the guerrillas to concentrate on their *military* [emphasis added] tasks.[10]

Furthermore, Brickhill quoted one ZIPRA regional commander who noted, "as far as controlling the local people was concerned, we left that to the civilians."[11]

Yet throughout the war, neither ZAPU nor ZIPRA emphasized the mass mobilization of the populace, as indicated by Cliffe, Mpofu, and Munslow:[12]

> So when the armed struggle was launched after UDI, it was hoped that the guerrillas would rely on the existing underground structures of ZAPU inside Zimbabwe. This precluded the necessity to mobilize and politicize the masses on the part of the guerrillas. They were deployed into a theatre of operation with instructions to "wage war against the white settler

189

regime" but little was said about politicizing and mobilizing the masses to participate in the war. The guerrillas were, however, instructed to make contact and work with the "underground officials of the party." The operative role of the said officials was: to organize and supply food, medicines and clothing for the guerrillas; to provide a courier-system upon which the guerrillas would rely for information and coordinating contacts; and to assist guerrillas in the recruitment of personnel for military training abroad. This pattern characterized ZIPRA guerrilla operations during the period 1965–1970. There was no politically substantiated program spelling out the objectives of armed struggle and, therefore, the politicization and preparation of the masses for a long drawn out liberation war was precluded. The absence of such political mobilization of the masses led to the total failure of the "strategy" of that period.

Furthermore, as the authors noted, a reevaluation of the war effort in 1970–1971 did not result in a greater emphasis on the large-scale mobilization of the populace:[13]

ZIPRA did not establish numerous operational structures characterized by coordinating committees and sub-committees on the scale that was undertaken by ZANLA forces in their base areas. The political aspect of their mission included "reviving" the "existing" structures and no reference was made to the political reorientation of the masses. The revitalized structures were simply to be organs for supplying medicines, uniforms, food and raw recruits. The courier system was established by the guerrillas themselves from specially selected persons, especially youths. The role of the couriers was different from that of the *mujibas* in ZANLA in that the formers' tasks included gathering of information about the enemy's activities in the ZIPRA operational areas. They did not possess any political power like the *mujibas*. They were merely messengers or intelligence operatives for ZIPRA. Without systematizing their political activities, ZIPRA cadres held political meetings with peasants in a somewhat crude fashion. People were called to meetings mostly to be given precautionary measures for maintaining security. Political meetings were called "once in many months."

The authors concluded:

> So through the bangs of bazookas and the rattling of gunfire, ZIPRA revived and revitalized not only the ZAPU – Patriotic Front Party but also the determination of the peasants to support the struggle. For the bangs were an absolute demonstration of power in physical terms which proved decisive for gaining and sustaining mass support in the operational areas. For ZIPRA there was no other way because they did not have a propaganda machinery nor a robust political commissariat to systematize political teaching in the population.[14]

Hence, a collapse of ZAPU's underground structure outside Matabeleland during the war, combined with an overall lack of emphasis in the mass mobilization of the population, appears to have hurt ZAPU/ Patriotic Front during the 1980 elections. What therefore can be said of Soviet training of ZAPU with respect to guerrilla tactics? As indicated in the previous chapter, Soviet training with respect to irregular and unconventional tactics appears to have emphasized the establishment of underground organizations along with carrying out ambushes and other forms of military training. As opposed to the training methods the Chinese used with ZANU, the Soviet training of ZAPU did not emphasize the mass mobilization of the population to resist the incumbent government in Salisbury. For Moorcraft and McLaughlin, "ZIPRA's guerrillas have been best described as 'mobile revolutionary vanguards,' operating much like Soviet partisans, rather than as Mao's type of guerrilla 'fish'."[15]

The preceding discussion largely reflects the historiography of the conflict, which emphasizes the importance of ZANU's people's war strategy in fostering and mobilizing peasant nationalism, which in turn led to electoral victory in the 1980 elections. However, there are several notable counterpoints to this argument. Previously, it was noted that ZIPRA did undertake population-centric measures, but through ZAPU. Interestingly, Moorcraft and McLaughlin noted:[16]

> Still, ZIPRA did wage a guerrilla campaign in western Rhodesia which was not dissimilar to the 'people's war' fought by ZANLA. . . . ZIPRA did not practice the mass politicization and night-time *pungwes* because it claimed that it had sufficient local support, nourished by the more extensive branch structure of the older ZAPU organization, although much of it had been driven underground.

More detail was provided by Sibanda, who sought to refute the prior argument presented by Cliffe, Mpofu, and Munslow by noting that:[17]

> Obviously, and contrary to the assertion of Cliffe, Mpofu and Munslow that ZIPRA did not establish an administrative network of the *umjibha*, committees and subcommittees, and that its *umjibha* did not possess any political power, the *umjibha* did possess political power, and his political activities were politically systematized, just as much as the activities of the local population were.

Further, he stated:[18]

> In so far as politicization was concerned, unlike his counterparts the ZANLA *umjibha*, who used all-night pungwes (rallies) to instill such information, ZIPRA umjimba addressed people in very small groups of no more than ten people at a time, excepting in church situations, and usually it was during the day. The idea was to avoid detection and situations in which Rhodesian soldiers killed the innocent unnecessarily. This was the approach of the ZIPRA forces, which tried by all means to avoid using civilians as human shields. The same could not be said about ZANLA, which addressed villagers in large groups, sometimes with calamitous results.

Addressing this topic, Brickhill noted that "ZIPRA '*mujibas*' played little role in political mobilization—except where this was a party activity."[19] While Sibanda made clear that ZIPRA *mujibas* engaged in the politicization of the populace, it is unclear whether this activity fell under the category of party activity and, even if it was, whether such activity can be characterized as a mass-based effort at politicization.

Another counterpoint was offered by Krieger, who noted that much of the existing historiography downplayed the role of the compulsion of the population by ZANU. Indeed, ZANLA guerrillas often used terror tactics against alleged "sellouts," as documented by Pandya:[20]

> 'Punishment' was usually carried out at a pungwe. When people reported to the guerrillas on those who were employed by the government, or those who were alleged to have 'collaborated' with the authorities, the accused were then put on 'trial' and invariably found guilty. The penalty was usually death. However, amputation of hands, fingers, legs, toes, lips, tongues and ears was regarded as a less severe form of 'punishment.' Depending on the judgment, 'punishment' was always carried out in public at the pungwe, for all to witness what happened to those who 'collaborated' with the authorities. This type of trial and 'punishment' instilled great fear amongst the rural people and they became very reluctant to associate themselves with the authorities in any way. This form of action consequently could play an extremely profound role in paralyzing contrary action or even the holding of opinions contrary to those subscribed by ZANU and ZANLA.

Above and beyond such atrocities, Krieger noted that ZANLA activities during the war imposed a number of costs on the Shona public. For instance, guerrilla demands for food and livestock were often perceived as onerous, and guerilla attacks on government infrastructure, such as schools, clinics, and transport facilities, deprived rural Shona residents of many services on which they had become reliant.[21] Citing the reluctance of individuals to serve on rural war committees, Krieger noted that rural support, especially among non-youths, was reluctant at best and that ZANU electoral support was based more on a desire to end the war and on the expectation that the party would compensate Shona peasants for the sacrifices ZANU imposed on them during the war.[22] The larger point is that Krieger's argument counters the dominant narrative that ZANU's electoral support naturally flowed from ZANU's adoption of a people's war military strategy.

Lastly, it is also important to consider how successful ZANLA's strategy of mass mobilization was in light of the ethnic divisions within Zimbabwe. Indeed, the importance of ethnicity in ZANLA's people's war strategy is highlighted by the fact that in 1972 ZAPU declined to act on FRELIMO's suggestion to launch operations into Rhodesia from the Tete province in Mozambique because the largely Ndebele guerrillas of ZAPU had little support from the Shona population in northeast Rhodesia. In blunt terms, the guerrillas would not have survived long in the northeast.[23] Additionally, ZANU (PF) did not poll well in regions where the Shona presence was minimal, which in turn raises the possibility that the success of ZANU (PF) at the polls simply reflects the efficacy of a strategy of ethnic mobilization rather than mass mobilization irrespective of the ethnic or tribal makeup of the populace. The question essentially boils down to whether the 1980 election results were a reflection more of tribalism rather than a people's war strategy per se. If the former, then a people's war strategy of building strong links between a guerrilla army and a populace, with the former mobilizing the latter, may be ineffective if a populace symbolically identifies with rival groups on the basis of ethnic and language affinities.

In this regard, Cliffe, Mpofu, and Munslow noted the experience of ZANLA guerrillas in Matabeleland South Province, where 60 percent of the populace spoke Ndebele in 1969.[24] ZANLA had established itself in about one-third of the province, yet ZANU (PF) did not win a single seat in the province during the 1980 elections. This outcome may be due to the clumsy and heavy-handed efforts of ZANLA guerrillas to appeal to the local populace, for instance by requiring people to chant "Pasi na Nkomo" (down with Nkomo) at rallies, as related by one participant:[25]

> When the ZANLA guerrillas arrived here, we welcomed them. We pledged our support to the cause they were fighting for. They demanded *sadza* (maize porridge); we gave them. They called us to their meetings; we attended. But we were shocked to be told to say, *Pasi na Nkomo*; denounce a man who has fought and suffered together with Muguabe for the freedom of our country? Is he not the founder of African

nationalism in Zimbabwe? Worse still: when we tried to resist saying the slogan, we became subjected to the most humiliating treatment ever. I cannot trust you to tell you all the details. Just take it from me that people here paid very heavily for refusing to denounce Nkomo. Maybe we will forget it now that the leaders are talking about reconciliation.

Such behavior earned ZANLA the moniker O-pasi (the down ones) from some inhabitants of the population. Additionally, some members objected to being addressed in Shona and questioned the arguments put forth by ZANLA members during efforts to politicize the population, as indicated in the following statement:[26]

I would have voted for ZANU-PF if O-pasi had treated me and my family properly. But they did not treat us in a good way. They tried to compel us to speak Shona. They were ruthless on those who asked why. All the slogans and songs were in Shona and they were not translated into Sindebele. Everything tended to "change" us to Shona. Yes, O-pasi told us that Nkomo had been shown to be unreliable by negotiating with Smith in 1976. They allege that he had a secret meeting with Smith in Lusaka without informing his comrade, Robert Mugabe. They said at the Lusaka meeting Nkomo wanted to give up the struggle if Smith promised him a high position in government. For doing this he was no longer fit to be a leader. That might be so, but then why did Mugabe and Nkomo remain co-leaders of the Patriotic Front Alliance if ZANU was unhappy with his way of doing this? At the top the leaders are working together but at the bottom we are told to denounce one of them. Why?

Such encounters led some members of the populace to falsify their political preferences publicly when dealing with ZANLA while covertly working with their electoral rival, the Patriotic Front headed by Nkomo:[27]

We gradually formed ourselves into small groups consisting of only those who had the same attitude against ZANLA slogans and treatment. When the election

campaign was launched, we quickly made covert contacts with PFP officials in Bulawayo and Gwanda.
We were issued with bundles of model ballot papers.
O-pasi too brought theirs and taught us how to vote
for Jongwe (cock)—the ZANU-PF election symbol. We
accepted both lots. The Njelele National rally pumped
new life in us. Mdala (old man) Nkomo made us feel
fresh and confident. After that rally the pro-PFP structures grew rapidly to cover all former ZANLA base
areas. To be safe we had to be double-faced. Of course
ZAPU blundered by not sending its guerrillas to cover
our area during [the] armed struggle. In any case if
ZANLA had treated us well, we would have had no reason to dislike them.

Such passages suggest that building links with ethnically and linguisti-
cally distinct populations may be possible if appeals based on shared
grievances are not delivered in a heavy-handed and culturally insensi-
tive manner.

NOTES

1   Ian Taylor, "Relations Between the PRC and Zimbabwe," *Issues and Studies* 33, no. 2 (1997): 125.

2   Justina Crabtree, "Zimbabwe Is Intent on 'Leapfrogging 18 Years of Isolation' with Chi-na's Help," *CNBC.com*, April 3, 2018, https://www.cnbc.com/2018/04/03/zimbabwes-president-emmerson-mnangagwa-makes-first-china-state-visit.html.

3   Jeremy Brickhill, "Daring to Storm the Heavens: The Military Strategy of ZAPU 1976 to 1979," in *Soldiers in Zimbabwe's Liberation War*, ed. Ngwabi Bhebe and Terence Ranger (London: James Currey, 1995), 70.

4   Martyn Gregory, "Zimbabwe 1980: Politicisation through Armed Struggle and Electoral Mobilisation," *Journal of Commonwealth and Comparative Politics* 19, no. 1 (1981): 74.

5   Ibid., 69–70.

6   Ken Flower, *Serving Secretly: An Intelligence Chief on Record Rhodesia into Zimbabwe 1964 to 1981* (London: John Murray Publishers Ltd., 1987), 110–111.

7   Jane Plastow, *African Theatre and Politics: The Evolution of Theatre in Ethiopia, Tanzania and Zimbabwe—A Comparative Study* (Amsterdam: Rodopi, 1996), 107–108.

8   Brickhill, "Daring to Storm the Heavens," *70–71.*

9   Ibid., 50–51.

10   Ibid., 69.

11   Ibid., 70.

[12] Lionel Cliffe, Joshua Mpofu, and Barry Munslow, "Nationalist Politics in Zimbabwe: The 1980 Elections and Beyond," *Review of African Political Economy* 7, no. 18 (1980): 55.

[13] Ibid., 56.

[14] Ibid., 57.

[15] Paul L. Moorcraft and Peter McLaughlin, *The Rhodesian War: A Military History* (Mechanicsburg, PA: Stackpole Books, 2008), 85.

[16] Ibid., 74.

[17] Eliakim M. Sibanda, *The Zimbabwe African People's Union 1961–87: A Political History of Insurgency in Southern Rhodesia* (Asmara, Eritrea: Africa World Press, Inc., 2005), 132.

[18] Ibid., 131.

[19] Brickhill, "Daring to Storm the Heavens," 70.

[20] Paresh Pandya, *Mao Tse-Tung and Chimurenga: An Investigation into ZANU's Strategies* (Johannesburg: Skotaville Educational Division, 1988), 173.

[21] Norma J, Krieger, *Zimbabwe's Guerrilla War: Peasant Voices* (Cambridge: Cambridge University Press, 1992), 144-147.

[22] Ibid., 161–165.

[23] Pandya, *Mao Tse-Tung and Chimurenga*, 78–79.

[24] Cliffe, Mpofu, and Munslow, "Nationalist Politics in Zimbabwe," 59.

[25] Ibid., 63.

[26] Ibid., 64.

[27] Ibid., 63.

APPENDICES

APPENDIX A. TIMELINE OF SIGNIFICANT EVENTS DURING THE RHODESIAN CONFLICT

December 17, 1961	The Zimbabwe African People's Union (ZAPU) forms under leadership of Joshua Nkomo.
1962	The Rhodesian Front (RF) party wins the election with Winston Field as prime minister.
September 1962	ZAPU is banned; the People's Caretaker Council (PCC) is created as a proxy replacement.
1963	ZAPU members dissatisfied with Nkomo's leadership form the Zimbabwe African National Union (ZANU). Ndabaningi Sithole is elected chairman and Robert Mugabe secretary. Emmerson Mnangagwa leads the first squad of recruits of the Zimbabwe African National Liberation Army (ZANLA, the military arm of ZANU) to China for guerrilla training.
December 31, 1963	The Central African Federation (CAF) disbands, relinquishing British control over Northern Rhodesia and Nyasaland.
1964	Pieter Oberholtzer becomes the first white killed in an act of war since the 1896–1897 Mashona Rebellion.
April 13, 1964	Winston Field is forced to resign over his reluctance to declare unilateral independence from Britain. His deputy, Ian Smith, assumes the position of prime minister.
June 22, 1964	ZANU and PCC are banned and Nkomo, Sithole, Mugabe, and other leading nationalists are arrested and imprisoned for the next ten years for their antigovernment activities.
July 6, 1964	Nyasaland gains independence, assumes majority rule, and becomes Malawi.
October 24, 1964	Northern Rhodesia assumes majority rule and becomes Zambia.
November 11, 1965	Ian Smith declares a state of emergency and issues the Unilateral Declaration of Independence (UDI) to Great Britain. The RF issues a new constitution that refers to the country simply as Rhodesia.

1965	With its leadership in prison, ZANU establishes itself at Lusaka, Zambia. Sithole directs Herbert Chitepo to assume the leadership of ZANU in his absence. ZAPU and ZANU compete for attention and resources.
1966	The United Nations (UN) imposes selective sanctions on Rhodesia.
April 28, 1966	Seven ZANLA guerrillas die in a battle at Sinoia. This is considered the beginning of the war for black nationalists, and the date is marked as Chimurenga Day, the start of the armed struggle for independence.
September 30, 1966	Botswana gains independence from Great Britain.
1967	Zimbabwe People's Revolutionary Army (ZIPRA) and South African ANC rebels infiltrate across the Zambezi River and engage Rhodesian Forces at Wankie. The operations (Operation Isotopes II and Operation Nickel) cause the Rhodesian government to reassess the guerrilla threat and change its counterinsurgency tactics.
1968	The UN imposes comprehensive mandatory sanctions on Rhodesia.
1969	Ndabaningi Sithole, the leader of ZANU, is sentenced to six years imprisonment for plotting to assassinate Ian Smith. During his trial he denounces armed violence, which sets in motion the event that will lead to his eventual loss of credibility and removal as ZANU's president in 1974.
	In November, ZANU leaders meet with FRELIMO to discuss establishing ZANLA bases in Mozambique.
1970	ZANLA and FRELIMO agree to allow ZANLA guerrillas to establish a base in Tete Province in Mozambique.
1971	Ian Smith and Britain's foreign secretary agree to a settlement proposal that would set a timetable for majority rule in Rhodesia. Nationalists oppose the agreement because it would likely keep white minority rule in place for decades. Bishop Muzorewa establishes the Africa National Council to oppose the proposals.
	ZANLA guerrillas, operating from their base in Tete, Mozambique, conduct their first scouting raids in Rhodesia.

1972	The Pearce Commission finds that the settlement proposal is unacceptable to the African majority (their agreement to accept the proposal was a condition of its implementation).
December 21, 1972	Sixty guerrillas infiltrate Rhodesia from Tete Province, Mozambique, in preparation for the beginning of the protracted ZANLA armed struggle. (This infiltration is known as the attack on Altena Farm.)
1973	Ian Smith closes the Zambian border after a dispute with the Zambian government. Despite Smith's declaration to reopen the border, Zambian President Kaunda keeps it closed for the duration of the war. Bishop Muzorewa starts negotiations with Smith.
1974	ZANLA operations from Mozambique escalate, leading the Rhodesian government to extend the length of conscripted service.
	Détente talks are called in Lusaka under the condition that key nationalist leaders be released from detention. Nkomo, Mugabe, and Sithole are released to attend the talks.
November 1974	UDI is signed by nationalist leaders but is undermined by the Nhari Rebellion (a revolt of ZANLA military members seeking better equipment and camp conditions).
April 25, 1974	A military coup in Portugal begins a series of events that lead to Mozambique gaining independence the following year. The secret defense alliance among Rhodesia, Mozambique and South Africa, known as the Alcora Exercise, collapses.
March 18, 1975	Herbert Chitepo, the acting commander of ZANU, is assassinated in Lusaka during an internal rebellion. Subsequently, Zambian authorities arrest ZANU senior members and ZANLA military leaders, disrupting the war effort for almost a year. As a result, ZANU and ZANLA forces are denied further access to training and staging bases in Zambia.
1975	Robert Mugabe assumes control of ZANU in defiance of Sithole's leadership and willingness to negotiate with the Rhodesian government.

June 25, 1975	Mozambique gains its independence from Portugal. FRELIMO assumes the seat of the new government. This dramatically changes the security situation along Rhodesia's eight-hundred-mile eastern border with Mozambique.
1976	The war resumes after a lull following the Nhari Rebellion and the aftermath of the Chitepo assassination. ZANU opens up three fronts from Mozambique: Tete, Manica, and Gaza. Rhodesian Security Forces (RSF) begin a concerted effort to destroy guerrilla bases in Mozambique. Mozambique closes its border with Rhodesia.
	US Secretary of State Kissinger tours Africa and joins South African President Vorster in persuading Ian Smith to accept the principle of majority rule in Rhodesia. Kaunda releases detained ZANU leaders.
1977	The United States rescinds the Chrome Import Law, depriving the Rhodesian government of critical foreign currency. South Africa softens its support to the Rhodesian government under pressure from the United States and the United Kingdom, hindering critical oil supplies.
	The economic toll of sanctions and growing requirement for resources to secure the country force Smith to seek a negotiated settlement.
	ZAPU leaders (Nkomo and Muzorewa) participate in negotiations while ZANU steps up its attacks.
November 23–25, 1977	RSF engage in highly successful cross border operation (Operation Dingo) against New Farm (Chimoio) and Tembue ZANLA camps in Mozambique. ZANLA suffers heavy casualties. The operation is widely portrayed internationally as an attack on a refugee camp.
1977	Robert Mugabe becomes the official leader of ZANU.
1978	Smith signs an agreement with Sithole and Muzorewa to establish a transition government preceding majority rule. ZAPU and ZANU leaders are nonsupportive. ZANLA vows to continue to fight.

June 1979	The White referendum approves the new Zimbabwe-Rhodesia constitution, but international recognition is withheld. New Prime Minster Bishop Muzorewa offers amnesty proposals to guerrillas, but they are ignored. Sanctions continue.
1979	The government under new British Prime Minister Thatcher calls for a conference in London including all parties. Mugabe is reluctant to attend the negotiations because he believes victory can and should be achieved militarily. Representatives of the Soviet Union persuade him to attend.
December 1979	The Lancaster House Agreement leads to a new Zimbabwean constitution.
February 1980	Robert Mugabe and the ZANU-Patriotic Front (PF) win the first Zimbabwean elections. Mugabe forms the first black government of Zimbabwe.

APPENDIX B. ACRONYMS

AAPSO	Afro-Asian People's Solidarity Organization
ANC	African National Congress
BSAP	British South African Police
CAF	Central African Federation
CAPFA	China African People's Friendship Association
CCP	Chinese Communist Party
CPSU	Communist Party of the Soviet Union
FRELIMO	Frente de Libertação de Moçambique (Mozambique Liberation Front)
FROLIZI	Front Line for the Liberation of Zimbabwe
ID	International Department
IDAF	International Defence and Aid Fund
MPLA	Movimento Popular de Libertação de Angola (People's Movement for the Liberation of Angola)
NCNA	New China News Agency
NDP	National Democratic Party
OAU	Organisation of African Unity
PAFMECSA	Pan African Movement of East, Central, and Southern Africa
PAIGC	Partido Africano da Independência da Guiné e Cabo Verde
PCC	People's Caretaker Council
PRC	People's Republic of China
PF	Patriotic Front
RF	Rhodesian Front
RSF	Rhodesian Security Forces
SFA	Security Force Auxiliaries
SRANC	Southern Rhodesia African National Congress
SWAPO	South West African People's Organization
TTL	Tribal Trust Lands
UANC	United African National Council
UDI	Unilateral Declaration of Independence
UN	United Nations
UNITA	National Union for the Total Independence of Angola
ZANLA	Zimbabwe African National Liberation Army
ZANU	Zimbabwe African National Union

ZAPU	Zimbabwe African People's Union
ZIPA	Zimbabwe People's Army
ZIPRA	Zimbabwe People's Revolutionary Army

BIBLIOGRAPHY

Albright, David E. "The Communist States and Southern Africa." In *International Politics and Southern Africa*, edited by Gwendolen M. Carter and Patrick O'Meara. Bloomington, IN: Indiana University Press, 1982.

Alden, Chris, and Ana Cristina Alves. "History & Identity in the Construction of China's Africa Policy." *Review of African Political Economy* 35, no. 115 (2008): 43–58.

Alexander, Jocelyn and JoAnn McGregor. "African Soldiers in the USSR: Oral Histories of ZAPU Intelligence Cadres' Soviet Training, 1964-1979." Journal of Southern African Studies 43, no. 1 (2017): 49–66.

Andrew, Christopher, and Vasili Mitrokhin. *The World Was Going Our Way: The KGB and the Battle for the Third World*. New York: Basic Books, 2005.

Astrow, Andre. *Zimbabwe: A Revolution That Lost Its Way?* London: Zed Press, 1983.

Barratt, John. *The Soviet Union and Southern Africa*. Braamfontein: The South African Institute of International Affairs, 1981.

Baxter, Peter. *Rhodesia, Last Outpost of the British Empire 1890–1980*. Alberton: Galago Publishing, 2010.

——— "ZAPU in the Zimbabwe Liberation Struggle." Part 19 of 20 of the "History of the amaNdebele" series. *Peter Baxter: Author, Speaker & Heritage Guide* (blog). Accessed October 14, 2015, http://peterbaxterafrica.com/index.php/2012/01/06/zapu-in-the-zimbabwe-liberation-struggle.

Bell, J. Bowyer. "The Frustration of Insurgency: The Rhodesian Example in the Sixties." *Military Affairs* 35, no. 1 (1971): 1–5.

Blake, Robert. *A History of Rhodesia*. New York: Alfred A. Knopf, Inc., 1977.

Bowman, Larry W. *Politics in Rhodesia: White Power in an African State*. Cambridge, MA: Harvard University Press, 1973.

Brazinsky, Gregg. Winning the Third World: Sino-American Rivalry During the Cold War. Chapel Hill: University of North Carolina Press, 2017.

Brickhill, Jeremy. "Daring to Storm the Heavens: The Military Strategy of ZAPU 1976 to 1979." In *Soldiers in Zimbabwe's Liberation War*, edited by Ngwabi Bhebe and Terence Ranger. London: James Currey, 1995.

British Broadcasting Company. "Immigration and Emigration: Zimbabwe – or Was it Rhodesia." *Legacies.* Accessed August 24, 2015, http://www.bbc.co.uk/legacies/immig_emig/england/gloucestershire/article_3.shtml.

Burgess, M. Elaine. "Ethnic Scale and Intensity: The Zimbabwean Experience." *Social Forces* 59, no. 3 (1981): 601–626.

Calvocoressi, Peter. *World Politics since 1945.* 9th ed. Abingdon, Oxon, England: Routledge, 2009.

Cary, Robert, and Diana Mitchell. *African Nationalist Leaders in Rhodesia – Who's Who.* Accessed October 12, 2015, originally published in 1977. http://www.colonialrelic.com.

Central Intelligence Agency. *Chinese Communist Activities in Africa.* Washington, DC: CIA Director of Intelligence, 1965.

Charlton, M. *The Last Colony in Africa: Diplomacy and the Independence of Rhodesia.* Oxford: Blackwell, 1990.

Chartrand, Philip E. "Political Change in Rhodesia: The South Africa Factor." *Issue: A Journal of Opinion* 5, no. 4 (1975): 13–20.

Chau, Donovan C. *Exploiting Africa: The Influence of Maoist China in Algeria, Ghana and Tanzania.* Annapolis: Naval Institute Press, 2014.

Chimurenga Day: 13th Anniversary of the Battle of Sinoia. Performed by Robert Mugabe. Maputo. April 28, 1979.

Chongo, Clarence. "Decolonising Southern Africa: A History of Zambia's Role in Zimbabwe's Liberation Struggle, 1964-1979." Ph.D. dissertation, University of Pretoria, 2015.

Chung, Fay. *Re-living the Second Chimurenga: Memories from the Liberation Struggle in Zimbabwe.* Stockholm: The Nordic Africa Institute, 2006.

Christensen, Thomas J. "Chinese Realpolitik: Reading Beijing's World-View." *Foreign Affairs* 75, no. 5 (September–October 1996).

Cilliers, J. K. *Counter-Insurgency in Rhodesia.* Dover, NH: Croom Helm Ltd., 1985.

Cliffe, Lionel, Joshua Mpofu, and Barry Munslow. "Nationalist Politics in Zimbabwe: The 1980 Elections and Beyond." *Review of African Political Economy* 7, no. 18 (1980): 44–67.

Crabtree, Justina. "Zimbabwe Is Intent on 'Leapfrogging 18 Years of Isolation' with China's Help." CNBC.com, April 3, 2018. Accessed at https://www.cnbc.com/2018/04/03/zimbabwes-president-emmerson-mnangagwa-makes-first-china-state-visit.html.

Dabengwa, Dumiso. "ZIPRA in the Zimbabwe War of National Liberation." In *Soldiers in Zimbabwe's Liberation War*, edited by Ngwabi Bhebe and Terence Ranger. London: James Currey, 1995.

Daniel, John. "Racism, the Cold War and South Africa's Regional Security Strategies 1948-1990" in ed. Sue Onslow. Cold War in Southern Africa: White Power, Black Liberation. London: Routledge, 2009.

Day, John. "The Divisions of the Rhodesian African Nationalist Movement." *The World Today* 33, no. 10 (1977): 385–394.

———. "The Rhodesian Internal Settlement." *The World Today* 34, no. 7 (1978): 268–276.

"Dr Dumiso Dabengwa Profile - Zimbabwe's liberator.wmv." YouTube video, 12:43. Posted by Zhou Media House, January 21, 2001. https://www.youtube.com/watch?v=UFHKgzI3gfA.

Ellert, H. *The Rhodesian Front War: Counterinsurgency and Guerrilla War in Rhodesia 1962–1980*. Gweru, Zimbabwe: Mambo Press, 1989.

Flower, Ken. *Serving Secretly: An Intelligence Chief on Record Rhodesia into Zimbabwe 1964 to 1981*. London: John Murray Publishers Ltd., 1987.

Friedman, N. *The Fifty-Year War: Conflict and Strategy in the Cold War*. Annapolis: Naval Institute Press, 2000.

Foster, Warren, Nosimilo Ndlovu, Zodidi Mhlana, and Surika Van Schalkwyk. "The 'Black Russian' Changes Sides." *Mail & Guardian*, March 6, 2008. http://mg.co.za/article/2008-03-06-the-black-russian-changes-sides.

Gann, Lewis H., and Thomas H. Henriksen. *The Struggle for Zimbabwe: Battle in the Bush*. New York: Praeger, 1981.

Gorman, Robert F. "Soviet Perspectives on the Prospects for Socialist Development in Africa." *African Affairs* 83, no. 331 (1984): 163–187.

Gregory, Martyn. "Zimbabwe 1980: Politicisation through Armed Struggle and Electoral Mobilisation." *Journal of Commonwealth and Comparative Politics* 19, no. 1 (1981): 63–94.

Gupta, Surendra, K. Stalin's Policy Towards India, 1946-1953. New Delhi: South Asian Publishers, 1988.

Halsall, Paul. "Rhodesia: Unilateral Declaration of Independence Documents, 1965." *Modern History Sourcebook.* July 1998. http://www.fordham.edu/halsall/mod/1965Rhodesia-UDI.html.

Hausmann, Ricardo, César A. Hidalgo, Sebastián Bustos, Michele Coscia, Sarah Chung, Juan Jimenez, Alexander Simoes, and Muhammed A. Yıldırım. *The Atlas of Economic Complexity.* Cambridge, MA: Puritan Press, 2011.

Hendrix, Cullen, and Idean Salehyan. "Zimbabwe." In *Civil Wars of the World: Major Conflicts since World War II,* edited by Karl DeRouen Jr. and Uk Heo Karl DeRouen, 829–846. Santa Barbara: ABC-CLIO, Inc., 2007.

Henkin, Yagil. "Stoning the Dogs: Guerilla Mobilization and Violence in Rhodesia." *Studies in Conflict & Terrorism* 36, no. 6 (2013): 503–532.

Herbstein, Denis. *White Lies: Canon Collins and the Secret War against Apartheid.* Oxford: James Currey, 2004.

Hoffman, Bruce, Jennifer Taw, and David Arnold. *Lessons for Contemporary Counterinsurgencies: The Rhodesian Experience.* Report No. R-3998-A. Santa Monica, CA: RAND Corporation, 1991.

Hunt, Michael H. The Genesis of Chinese Communist Foreign Policy. New York: Columbia University Press, 1996.

Jacobson, Jon. When the Soviet Union Entered World Politics. Berkely, CA: University of California Press, 1994.

Jackson, Steven F. "China's Third World Foreign Policy: The Case of Angola and Mozambique, 1961–93." *The China Quarterly,* no. 142 (1995): 388–422.

Jayawardena, Amal. "Soviet Involvement in South Asia: The Security Dilemma." In *Security Dilemma of a Small State: Sri Lanka in the South Asian Context,* edited by P. V. J. Jayasekera. New Delhi: South Asian Publishers Pvt. Ltd., 1992.

Jian, Chen. *Mao's China and the Cold War.* Chapel Hill: University of North Carolina Press, 2001.

Johnson, R. W. "How Mugabe Came to Power: R.W. Johnson Talks to Wilfred Mhanda." *London Review of Books* 23, no. 4 (2011): 26–27.

Kandiah, Michael, and Sue Onslow, eds. *Britain and Rhodesia: The Road to Settlement.* London: Institute of Contemporary British History, 2008.

Kempton, Daniel R. *Soviet Strategy toward Southern Africa: The National Liberation Movement Connection.* New York: Praeger, 1989.

Khadiagala, Gilbert M. Allies in Adversity: The Frontline States in Southern African Security, 1975-1993. Athens, Ohio: Ohio University Press, 1994.

Kramer, Mark. "The Role of the CPSU International Department in Soviet Foreign Relations and National Security Policy." *Soviet Studies* 42, no. 3 (1990): 429–446.

Krieger, Norma J. *Zimbabwe's Guerrilla War: Peasant Voices.* Cambridge: Cambridge University Press, 1992.

Lan, David. *Guns and Rain: Guerrillas and Spirit Mediums in Zimbabwe.* Berkeley, CA: University of California Press, 1985.

Larkin, Bruce D. *China and Africa, 1949–1970.* Berkeley, CA: University of California Press, 1971.

Legum, Colin. "The Soviet Union, China, and the West in Southern Africa." *Foreign Affairs* 54, no. 4 (July 1976): 745–762.

Leonard, Richard. South Africa at War: White Power and the Crisis in Southern Africa. Westport, Connecticut: Lawrence Hill & Company, 1983.

Lenin, Vladimir I. "Report on the Unity Congress of the R.S.D.L.P." In *Lenin Collected Works, vol. 10.* Moscow: Progress Publishers, 1962.

Lenman, B. *Chambers Dictionary of World History.* Edinburgh: Chambers, 2000.

Mavhunga, Clapperton Chakanetsa. "A Plundering Tiger with its Deadly Cubs? The USSR and China as Weapons in the Engeering of a 'Zimbabwe Nation,' 1945-2009" in Entangled Geographies: Empire and Technopolitics in the Global Cold War. Cambridge, MA: MIT Press, 2011.

Mao Tse-tung. "Problems of War and Strategy." In *Selected Works of Mao Tse-tung: Vol. II.* Peking, China: Foreign Languages Press, 1938.

McKinnell, Robert. "Sanctions and the Rhodesian Economy." *Journal of Modern African Studies* 7, no. 4 (1969): 559–581.

Meisenhelder, Tom. "The Decline of Socialism in Zimbabwe." *Social Justice* 21, no. 4 (1994): 83–101.

"'Mhanda Not a National Hero' Says Zanu-PF." *Harare24 News*, May 31, 2014. http://harare24.com/index-id-news-zk-20797.html.

Martin, David, and Phyllis Johnson. *The Struggle for Zimbabwe*. London: Faber and Faber, 1981.

McWhirter, William. "World: This War Must End." *Time*, January 14, 1980.

Ministry of Information. *The Murder of Missionaries in Rhodesia*. Rhodesia: Ministry of Information, 1978. http://www.archive.org/stream/TheMurderOfMissionariesInRhodesia/MMR_djvu.txt.

Minter, William, and Elizabeth Schmidt. "When Sanctions Worked: The Case of Rhodesia Reexamined." *African Affairs* 87, no. 347 (1988): 207–237.

Mlambo, Alois. "Building a White Man's Country: Aspects of White Immigration into Rhodesia up to World War II." *Zambezia* 25, no. 2 (1998): 123–146.

———. "From the Second World War to UDI, 1940–1965." In *Becoming Zimbabwe*, edited by Brian Raftopoulos and Alois Mlambo, 75–114. Harare: Weaver Press, 2009.

———. "'Some Are More White Than Others': Racial Chauvinism as a Factor in Rhodesian Immigration Policy, 1890–1963." *Zambezia* 27, no. 2 (2000): 139–160.

Moorcraft, Paul L., and Peter McLaughlin. *The Rhodesian War: A Military History*. Mechanicsburg, PA: Stackpole Books, 2008.

Moorcraft, Paul. "Rhodesia's War of Independence." *History Today* 40, no. 9 (September 1990).

Morapedi, Wazha G. "The Dilemmas of Liberation in Southern Africa: The Case of Zimbabwean Liberation Movements and Botswana, 1960-1979." Journal of Southern African Studies 38, no. 1 (2012): 73-90.

Mosia, Lebona, Charles Riddle, and Jim Zaffiro. "From Revolutionary to Regime Radio: Three Decades of Nationalist Broadcasting in Southern Africa." *Africa Media Review* 8, no. 1 (1994): 1–24.

Natufe, O. Igho. *Soviet Policy in Africa: From Lenin to Brezhnev*. Bloomington, IN: iUniverse, Inc., 2011.

Ndlovu-Gatsheni, Sabelo J. "Angola–Zimbabwe Relations: A Study in the Search for Regional Alliances." *The Round Table* 99, no. 411 (2010): 631–653.

————. "Nationalist-Military Alliance and the Fate of Democracy in Zimbabwe." *African Journal of Conflict Resolution* 6, no. 1 (2006): 49–80.

Nkala, Thulani. "Happy Birthday Lookout Masuku: Retracing His Footsteps." *Harare24 News*, April 6, 2012. http://harare24.com/index-id-Opinion-zk-13958.html.

Novak, Andrew. "Abuse of State Power: The Mandatory Death Penalty for Political Crimes in Southern Rhodesia, 1963–1970." *Fundamina* 19, no. 1 (2013): 28–47.

Ogunsanwo, Alaba. *China's Policy in Africa, 1958–71.* London: Cambridge University Press, 1974.

Ojakorotu, Victor, and Ayo Whehto. "Sino-African Relations: The Cold War Years and After." *Asia Journal of Global Studies* 2, no. 2 (2008): 35–42.

Olden, Mark. "This Man Has Been Called Zimbabwe's Che Guevara. Did Mugabe Have Him Murdered?" *The New Statesman* (UK), April 12, 2004. http://www.newstatesman.com/node/195000.

Onslow, Sue. "The Cold War in Southern Africa: White Power, Black Nationalism and External Intervention." In *Cold War in Southern Africa: White Power, Black Liberation,* edited by Sue Onslow. London and New York: Routledge, 2009.

Palmer, R. *Land and Racial Discrimination in Rhodesia.* Berkeley, CA: University of California Press, 1977.

Pandya, Paresh. Mao Tse-Tung and Chimurenga: An Investigation into ZANU's Strategies. Johannesburg: Skotaville Educational Division, 1988.

Phiri, Bizeck Jube. "The Capricorn Africa Society Revisited: The Impact of Liberalism in Zambia's Colonial History, 1949–1963." *International Journal of African Historical Studies* 24, no. 1 (1991): 65–83.

Plastow, Jane. *African Theatre and Politics: The Evolution of Theatre in Ethiopia, Tanzania and Zimbabwe—A Comparative Study.* Amsterdam: Rodopi, 1996.

Pratt, R. Cranford. "Partnership and Consent: The Monckton Report Reexamined." *International Journal* 16, no. 1 (1960/1961): 37–49.

Powell, Nathaniel Kinsey. "The UNHCR and Zimbabwean Refugees in Mozambique, 1975–1980." *Refugee Survey Quarterly* 32, no. 4 (2013): 41–65.

Raeburn, Michael. *We Are Everywhere: Narratives from Rhodesian Guerrillas.* New York: Random House, 1978.

Reed, William Cyrus. "International Politics and National Liberation: ZANU and the Politics of Contested Sovereignty." *African Studies Review* 36, no. 2 (1993): 31–59.

Reuters. "Lookout Masuku Dies at 46; Commanded Nkomo Forces." *New York Times*, April 7, 1986.

Rhodesian Ministry of Foreign Affairs. *Communist Support and Assistance to Nationalist Political Groups in Rhodesia.* Rhodesia: Information Section, Ministry of Foreign Affairs, November 28, 1975. Posted on *Rhodesia and South Africa: Military History* (blog). Accessed August 22, 2014, http://www.rhodesia.nl/commsupp.htm.

Richelson, Jeffrey T. *Foreign Intelligence Organizations.* Cambridge, MA: Ballinger Publishing Company, 1988.

The Role of the Soviet Union, Cuba, and East Germany in Fomenting Terrorism in Southern Africa: Hearings Before the Subcommittee on Security and Terrorism of the Committee on the Judiciary. United States Senate, 97th Cong., 2d sess., 1982.

Scarnecchia, Timothy. *The Urban Roots of Democracy and Political Violence in Zimbabwe.* Rochester: University of Rochester Press, 2008.

Shamuyarira, Nathan. *Crisis in Rhodesia.* New York.: Transatlantic Arts, 1965.

Shinn, David H. "China's Involvement in Mozambique." *International Policy Digest,* August 2, 2012. http://www.internationalpolicydigest.org/2012/08/02/chinas-involvement-in-mozambique.

Shinn, David H., and Joshua Eisenman. *China and Africa: A Century of Engagement.* Philadelphia: University of Pennsylvania Press, 2012.

Shultz, Richard H. *The Soviet Union and Revolutionary Warfare: Principles, Practices and Regional Comparisons.* Stanford: Hoover Institution Press, 1988.

Shubin, Vladimir. *The Hot 'Cold War': The USSR in Southern Africa.* London: Pluto Press, 2008.

———— "Unsung Heroes: The Soviet Military and the Liberation of Southern Africa." *Cold War History* 7, no. 2 (2007): 251–262.

———— "Moscow and Zimbabwe's Liberation." Journal of Southern African Studies 43, no. 1 (2017): 225–233.

Sibanda, Eliakim M. *The Zimbabwe African People's Union 1961–1987: A Political History of Insurgency in Southern Rhodesia.* Asmara, Eritrea: Africa World Press, Inc., 2005.

Sjollema, Baldwin. *Never Bow to Racism: A Personal Account of the Ecumenical Struggle.* Geneva: World Council of Churches, 2015.

Smith, Ian. Bitter Harvest: Zimbabwe and the Aftermath of Independence: The Memoirs of Africa's Most Controversial Leader. London: John Blake, 2008.

Somerville, Keith. "The U.S.S.R. and Southern Africa Since 1976." *Journal of Modern African Studies* 22, no. 1 (1984): 73–108.

Southern Rhodesia: Report of the Constitutional Conference Held at Lancaster House, September–December 1979. London: Her Majesty's Stationery Office, 1980.

Spence, J. E. "Southern Africa in the Cold War." *History Today* 49, no. 2 (February 1999).

Starr, S. Frederick. *Xinjiang: China's Muslim Borderland.* Armonk, NY: M.E. Sharpe, Inc., 2004.

Stephenson, Glenn V. "The Impact of International Economic Sanctions on the Internal Viability of Rhodesia." *Geographical Review* 65, no. 3 (1975): 377–389.

Stewart, Michael A. *The Rhodesian African Rifles: The Growth and Adaptation of a Multicultural Regiment through the Rhodesian Bush War, 1965–1980.* Art of War Papers. Fort Leavenworth, KS: Combat Studies Institute Press, 2012.

Tavuyanago, Baxter. "The 'Crocodile Gang' Operation: A Critical Reflection on the Genesis of the Second Chimurenga in Zimbabwe." *Global Journal of Human Social Science* 13, no. 4 (2013): 27–36.

Taylor, Ian. "Relations Between the PRC and Zimbabwe." Issues and Studies 33, no. 2 (1997): 125–144.

Taylor, Ian. *China and Africa: Engagement and Compromise.* London: Routledge, 2006.

Tekle, Amare. "A Tale of Three Cities: The OAU and the Dialectics of Decolonization in Africa." *Africa Today* 35, no. 3/4 (1988): 49–60.

Thom, William. "Trends in Soviet Support for African Liberation." *Air University Review* (July–August 1974).

Thompson, W. Scott. *Ghana's Foreign Policy 1957–1966.* Princeton, NJ: Princeton University Press, 1969.

Tribute to Retired General Comrade Solomon Mujuru aka Comrade Rex Nhongo RIP (blog). August 16, 2011. http://solomonmujuru.blogspot.com.

Tse-Tung, Mao. *On Guerrilla Warfare.* Translated by Samuel B. Griffith II. Urbana: University of Illinois Press, 2000.

Tungamirai, Josiah. "Recruitment to ZANLA: Building up a War Machine." In *Soldiers in Zimbabwe's Liberation War,* edited by Ngwabi Bhebe and Terence Ranger. London: James Currey, 1995.

Ulam, Adam B. *Expansion and Coexistence: Soviet Foreign Policy 1917-1973* 2nd Edition. New York: Holt, Rinehart and Winston, Inc., 1974.

United Nations. *Decolonization No 5, Southern Rhodesia.* Geneva: United Nations Department of Political Affairs, Trusteeship and Decolonization, 1975.

United States Senate Subcommittee on Security and Terrorism. "The Role of the Soviet Union, Cuba, and East Germany in Fomenting Terrorism in Southern Africa." Washington, DC: US GPO, 1982.

Valentino, Benjamin, Paul Huth, and Dylan Balch-Lindsay. "'Draining the Sea': Mass Killing and Guerrilla Warfare." *International Organization* 58, no. 2 (2004): 375–407.

Vo Nguyen Giap. *People's War, People's Army.* Hanoi: Foreign Languages Press, 1961.

Watts, Carl Peter. "The 'Wind of Change': British Decolonisation in Africa, 1957–65." *History Review* 71 (December 2011): 12–17.

White, Luise. *The Assassination of Herbert Chitepo: Texts and Politics in Zimbabwe.* Bloomington, IN: Indiana University Press, 2003.

White, Luise. "'Normal Political Activities': Rhodesia, the Pearce Commission, and the African National Council." *Journal of African History* 52 (2011): 321–340.

Wilkinson, Anthony R. *Insurgency in Rhodesia, 1957–1973: An Account and Assessment.* Adelphi Paper No. 100. London: The International Institute for Strategic Studies, 1973.

Wilson, Edward T. *Russia and Black Africa Before World War II.* New York and London: Holmes & Meier Publishers, 1974.

Wood, J. R. T. "Rhodesian Insurgency." Rhodesia & South Africa website. Accessed November 10, 2014, http://www.rhodesia.nl/wood1.htm.

Yu, George T. "China's Role in Africa." *Annals of the American Academy of Political and Social Science* 432, no. 1 (1977): 96–109.

Zagoria, Donald. *The Sino-Soviet Conflict, 1956–1961.* Princeton, NJ: Princeton University Press, 1962.

Zickel, Raymond. *Soviet Union (Former): A Country Study.* Washington, DC: Library of Congress, 1989.

"Zimbabwe's Liberation A Short and Accurate History." YouTube video, 47:20. Posted by Antar Gholar, February 26, 2014. https://www.youtube.com/watch?v=z7iCjZf8ZGw.

" 'Zimbabwe, We Love You', As the Rebels Stream in from the Bush, Only Scattered Violence Mars the Truce." *Time,* January 14, 1980.

INDEX

A

225

F

www.ingramcontent.com/pod-product-compliance
Lightning Source LLC
Chambersburg PA
CBHW052110020426

42335CB00021B/2702